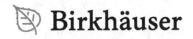
Birkhäuser

Applied and Numerical Harmonic Analysis

Lecture Notes in Applied and Numerical Harmonic Analysis

More information about this subseries at http://www.springer.com/series/13412

Shaun Ault · Charles Kicey

Counting Lattice Paths Using Fourier Methods

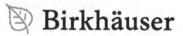 Birkhäuser

Shaun Ault
Department of Mathematics
Valdosta State University
Valdosta, GA, USA

Charles Kicey
Department of Mathematics
Valdosta State University
Valdosta, GA, USA

ISSN 2296-5009 ISSN 2296-5017 (electronic)
Applied and Numerical Harmonic Analysis
ISSN 2512-6482 ISSN 2512-7209 (electronic)
Lecture Notes in Applied and Numerical Harmonic Analysis
ISBN 978-3-030-26695-0 ISBN 978-3-030-26696-7 (eBook)
https://doi.org/10.1007/978-3-030-26696-7

Mathematics Subject Classification (2010): 05A10, 05A15, 42A16

This book is published under the imprint Birkhäuser, www.birkhauser-science.com by the registered company Springer Nature Switzerland AG
The registered company address is: Gewerbestrasse 11, 6330 Cham, Switzerland

LN-ANHA Series Preface

The *Lecture Notes in Applied and Numerical Harmonic Analysis (LN-ANHA)* book series is a subseries of the widely known *Applied and Numerical Harmonic Analysis (ANHA)* series. The Lecture Notes series publishes paperback volumes, ranging from 80 to 200 pages in harmonic analysis as well as in engineering and scientific subjects having a significant harmonic analysis component. *LN-ANHA* provides a means of distributing brief-yet-rigorous works on similar subjects as the *ANHA* series in a timely fashion, reflecting the most current research in this rapidly evolving field.

The *ANHA* book series aims to provide the engineering, mathematical, and scientific communities with significant developments in harmonic analysis, ranging from abstract harmonic analysis to basic applications. The title of the series reflects the importance of applications and numerical implementation, but richness and relevance of applications and implementation depend fundamentally on the structure and depth of theoretical underpinnings. Thus, from our point of view, the interleaving of theory and applications and their creative symbiotic evolution is axiomatic.

Harmonic analysis is a wellspring of ideas and applicability that has flourished, developed, and deepened over time within many disciplines and by means of creative cross-fertilization with diverse areas. The intricate and fundamental relationship between harmonic analysis and fields such as signal processing, partial differential equations (PDEs), and image processing is reflected in our state-of-the-art *ANHA* series.

Our vision of modem harmonic analysis includes mathematical areas such as wavelet theory, Banach algebras, classical Fourier analysis, time-frequency analysis, and fractal geometry, as well as the diverse topics that impinge on them.

For example, wavelet theory can be considered an appropriate tool to deal with some basic problems in digital signal processing, speech and image processing, geophysics, pattern recognition, bio-medical engineering, and turbulence. These areas implement the latest technology from sampling methods on surfaces to fast algorithms and computer vision methods. The underlying mathematics of wavelet theory depends not only on classical Fourier analysis but also on ideas from abstract harmonic analysis, including von Neumann algebras and the affine group. This leads to a study of the Heisenberg group and its relationship to Gabor systems and of the metaplectic group for a meaningful interaction of signal decomposition methods.

The unifying influence of wavelet theory in the aforementioned topics illustrates the justification for providing a means for centralizing and disseminating information from the broader, but still focused, area of harmonic analysis. This will be a key role of *ANHA*. We intend to publish with the scope and interaction that such a host of issues demands.

Along with our commitment to publish mathematically significant works at the frontiers of harmonic analysis, we have a comparably strong commitment to publish major advances in applicable topics such as the following, where harmonic analysis plays a substantial role:

Bio-mathematics, bio-engineering, and bio-medical signal processing;
Communications and RADAR;
Compressive sensing (sampling) and sparse representations;
Data science, data mining and dimension reduction;
Fast algorithms;
Frame theory and noise reduction;

Image processing and super-resolution;
Machine learning;
Phaseless reconstruction;
Quantum informatics;
Remote sensing;
Sampling theory;
Spectral estimation;
Time-frequency and time-scale analysis
– Gabor theory and wavelet theory

The above point of view for the *ANHA* book series is inspired by the history of Fourier analysis itself, whose tentacles reach into so many fields.

In the last two centuries Fourier analysis has had a major impact on the development of mathematics, on the understanding of many engineering and scientific phenomena, and on the solution of some of the most important problems in mathematics and the sciences. Historically, Fourier series were developed in the analysis of some of the classical PDEs of mathematical physics; these series were used to solve such equations. In order to understand Fourier series and the kinds of solutions they could represent, some of the most basic notions of analysis were defined, for example, the concept of "function." Since the coefficients of Fourier series are integrals, it is no surprise that Riemann integrals were conceived to deal with uniqueness properties of trigonometric series. Cantor's set theory was also developed because of such uniqueness questions.

A basic problem in Fourier analysis is to show how complicated phenomena, such as sound waves, can be described in terms of elementary harmonics. There are two aspects of this problem: first, to find, or even define properly, the harmonics or spectrum of a given phenomenon, e.g., the spectroscopy problem in optics; second, to determine which phenomena can be constructed from given classes of harmonics, as done, for example, by the mechanical synthesizers in tidal analysis.

Fourier analysis is also the natural setting for many other problems in engineering, mathematics, and the sciences. For example, Wiener's Tauberian theorem in Fourier analysis not only characterizes the behavior of the prime numbers but also provides the proper notion of spectrum for phenomena such as white light; this latter process leads to the Fourier analysis associated with correlation functions in filtering and prediction problems, and these problems, in turn, deal naturally with Hardy spaces in the theory of complex variables.

Nowadays, some of the theory of PDEs has given way to the study of Fourier integral operators. Problems in antenna theory are studied in terms of unimodular trigonometric polynomials. Applications of Fourier analysis abound in signal processing, whether with the fast Fourier transform (FFT), or filter design, or the adaptive modeling inherent in time-frequency-scale methods such as wavelet theory.

The coherent states of mathematical physics are translated and modulated Fourier transforms, and these are used, in conjunction with the uncertainty principle, for dealing with signal reconstruction in communications theory. We are back to the raison d'être of the *ANHA* series!

University of Maryland   John J. Benedetto
College Park, MD, USA   Series Editor

Preface

The primary topic of this monograph—counting certain types of lattice paths—seems far removed from the research interests of both authors (Ault has worked mainly in algebraic topology up to this point, while Kicey began his career in functional analysis, though now considers himself a generalist); however both mathematicians are fascinated by problems in enumerative combinatorics. While studying a problem called the *Sharing Problem*, Kicey, Katheryn Klimko, and Glen Whitehead [35] noticed familiar sequences in the ranges of the so-called *circular Pascal arrays*. To make a long story short, they, along with another of Kicey's students, Jonathon Bryant, discovered sequences of powers of two, the famous and ubiquitous Fibonacci sequence, and other sequences that were not immediately recognizable. When Ault was first introduced to these sequences, he suggested locating them in the *Online Encyclopedia of Integer Sequences* (OEIS). Sure enough, the first few cases were found. The entry for one case in particular (A061551) gives a tantalizing clue that all of these number sequences are indeed related. The main description of A061551 is: "number of paths along a corridor width 8, starting from one side," and further down the page, the following note can be found [54].

> Narrower corridors effectively produce A000007, A000012, A016116, A000045, A038754, A028495, A030436. An infinitely wide corridor (i.e., just one wall) would produce A001405.

Some of these sequences are well known. For example, A016116 consists of powers of two (repeated in pairs), A000045 is the Fibonacci sequence, and the numbers comprising A001405 are better known as the *central binomial coefficients*. Thenceforth we began a study in earnest of these sequences of **corridor numbers**, which solve certain types of lattice path counting problems. The authors' major insight in [6] was the introduction of the **dual corridor** structure, which helped to prove the link between corridor numbers and the circular Pascal array. In retrospect, we realized that the dual corridor structure encodes the effect of André's *reflection principle* [3] and associated *inclusion-exclusion* arguments.

The literature is replete with material on the subject of lattice path enumeration, including well-known connections to binomial coefficients, Catalan numbers, and other combinatorial objects of interest. For example, corridor numbers are useful in graph theory for counting the number of paths in the **path graph** P_m that start at a specified node. The interested reader may find a great deal written about lattice path combinatorics (e.g., [4, 17, 26, 33, 36, 37, 43, 45]). So what do we have to add to the discussion—especially coming from such diverse fields of research not immediately connected to enumerative combinatorics? Most of the formulae we develop, though rather general, are not new. Most have already been developed or implied using the traditional techniques of analyzing *recursively defined functions*, building *(ordinary) generating functions*, studying *continued fractions*, making use of special *inversion formulae*, and employing many other powerful tools. While we acknowledge and thoroughly respect the hard work done over many decades, going all the way back to Bertrand, André, and Lagrange in the 1800s, we would like to discuss yet another method for analyzing and counting lattice paths.

In fact, our methods are not part of a mainstream research program in enumerative combinatorics. Instead, they arose in part from working with talented undergraduates using familiar methods from elementary mathematical analysis and complex variables—more specifically, using the **discrete Fourier transform (DFT)** as a kind of periodic generating function. The fact that we use Fourier techniques and related aspects of mathematical analysis may suggest that our work has to do with Harmonic analysis. However, because we are addressing problems that generally fall outside the purview of Harmonic analysis (counting lattice paths), this work does not neatly fit into this category either. On the other hand, the history of mathematics is peppered with instances in which two diverse fields of study have merged to solve a problem resulting in a beautifully cohesive theory that transcends both. We find that our analytic approach unifies many different types of lattice path combinatorics results, and we hope that you will agree that our methods are compelling and useful. Furthermore, we hope that you find that our treatment is accessible to an undergraduate math major or minor (all that is required is a standard calculus sequence including complex functions and elementary linear algebra); in fact, we recommend this book for an undergraduate seminar or REU program. There are a number of challenging exercises at the end of each section, with selected solutions posted in the back of the book. Following each exercise section, we have listed a number of research questions that may serve as starting points for further work.

Acknowledgements We would like to acknowledge the support of the Department of Mathematics at Valdosta State University. We also appreciate the time and work put in by the various reviewers of this text to help us to clarify, focus, and refine our exposition. Shaun would like to thank his wife Megan and children Joshua, Holley, Samuel, and Felix, as constant sources of inspiration. Charles would like to thank his wife Paula for her patience, his parents Ed and Rose, and Chris Lennard for instilling the excitement of mathematics many years ago.

Valdosta, GA, USA   Shaun Ault
April 2019   Charles Kicey

Contents

Chapter 1
Lattice Paths and Corridors

In this chapter, we define our main combinatorial object of interest: corridor paths, which are a type of lattice paths. We introduce the vertex numbers as functions defining the numbers of such paths ending at a given point, and provide motivation for the techniques of the next chapter. The discussion herein is limited to combinatorics; Fourier techniques will not come into play until Chap. 2.

1.1   Corridor Paths

Throughout this text we denote the sets of natural numbers, integers, real numbers, and complex numbers by $\mathbb{N}$, $\mathbb{Z}$, $\mathbb{R}$, and $\mathbb{C}$ respectively, and let $\mathbb{N}_0 = \{n \in \mathbb{Z} : n \geq 0\}$. Initially, we define a **lattice** to be the integer-component points $\mathbb{Z}^r \subset \mathbb{R}^r$ for some $r \geq 1$, or a subset of $\mathbb{Z}^r$. A **lattice path** is a sequence of points of a lattice following prescribed rules (e.g., adjacent points of the path must be close to one another in the lattice, and moves can only happen in certain directions, etc.). Lattice path enumeration has wide-ranging applications, including random walks, parenthesis matching, tree structures, queues, and many others.[1]

The idea of a lattice path may find its origin in the famous *Ballot Problem* of Bertrand [10]:

> Suppose that n voters vote for either candidate A or candidate B. What is the probability as the votes are tallied one by one, that candidate A was never behind candidate B?

If a vote for candidate A corresponds to an up-right move in the lattice $\mathbb{Z}^2$, and a vote for candidate B corresponds to a down-right move, then the problem can be phrased in terms of lattice paths, by asking how many of the unrestricted lattice paths do not fall below the x-axis? For example, see Fig. 1.1.

---

[1]For background and applications, see [24, 26, 43].

© Springer Nature Switzerland AG 2019
S. Ault and C. Kicey, *Counting Lattice Paths Using Fourier Methods*, Applied and Numerical Harmonic Analysis,
https://doi.org/10.1007/978-3-030-26696-7_1

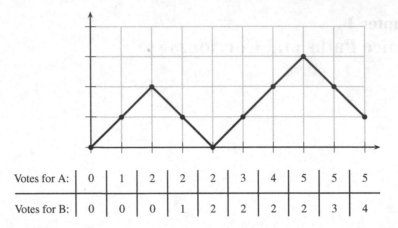

Votes for A:	0	1	2	2	2	3	4	5	5	5
Votes for B:	0	0	0	1	2	2	2	2	3	4

Fig. 1.1 Bertrand's Ballot Problem may be solved by counting certain kinds of lattice paths. A particular election consists of votes for candidate A (up-right) and B (down-right) in some sequence. A particular lattice path records the sequence of votes in one run of the election. If the path never dips below the x-axis, then that path represents an election in which candidate A was never behind B in the vote count

In this chapter we introduce the idea of *corridor paths* as an umbrella term covering various related kinds of lattice paths. Our main object of study is a restricted lattice called a **corridor**, which is a rectangular region within the lattice $\mathbb{N}_0 \times \mathbb{Z}^r$, for a fixed $r \in \mathbb{N}$. Let's begin with the simplest case, what we term the *one-dimensional corridor* ($r = 1$).

Definition 1.1 Let $h \in \mathbb{N}$, and let $\mathcal{C}^{(h)} \subset \mathbb{Z}$ be the set of lattice points $\{1, 2, \ldots, h\}$. The (h)-**corridor** is the set $\mathbb{N}_0 \times \mathcal{C}^{(h)}$. $\mathcal{C}^{(h)}$ is called the **fundamental region** for the (h)-corridor, and h is called the **height** of the corridor.

Remark 1.1 The fact that our fundamental regions start at 1 rather than 0 may seem arbitrary, but this choice helps to simplify formulas later on. In particular, the *dual corridor structure* introduced in Sect. 1.4 and the Fourier techniques of Chaps. 2 and 3 become more natural in this context.

Definition 1.2 A (**two-way**) **corridor path** is a lattice path in the (h)-corridor satisfying the following rules:

1. The initial point of the path is at $(0, a)$ for some fixed $a \in \mathcal{C}^{(h)}$.
2. The path never leaves the corridor.
3. Each step in the path is either an up-right or down-right move.

If a corridor path terminates at (n, k), then we say that the path has length n. There is one path of length zero, namely the path that begins and ends at $(0, a)$. If we regard the variable $n \in \mathbb{N}_0$ as (discrete) *time* and $k \in \{1, 2, \ldots, h\}$ as *position*,

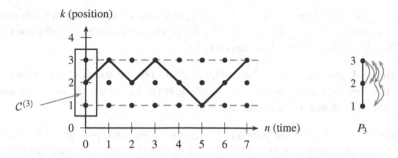

Fig. 1.2 *(Left)* A path of length 7 in the (3)-corridor with starting point $a = 2$. *(Right)* Path graph P_3, which is a graph whose vertex set is the fundamental region $\mathcal{C}^{(3)}$, and corresponding walk, 2, 3, 2, 3, 2, 1, 2, 3

then a corridor path may also be identified as a walk in the path graph[2] P_h starting at vertex a (where the vertices of P_h are identified with the points of $\mathcal{C}^{(h)}$). An example of a corridor path, along with its interpretation as a walk, is shown in Fig. 1.2.

In probability theory, one can interpret a lattice path as a *drunkard's walk* [24], in which a drunkard aimlessly walks along an infinite street sometimes walking forward and sometimes backward with equal probability. Corridor paths can then be considered as restricted drunkard's walks.[3] These types of lattice paths, their variations, and applications have been studied extensively in the combinatorial literature.[4] The purpose of this book is not to survey existing methods for lattice path enumeration, but instead to introduce a novel way of counting various kinds of lattice paths based on discrete Fourier methods.[5] Toward that end, we presently formulate the problem in terms of functions that we call *state vectors*.

1.2 Corridor Numbers and State Vectors

Throughout this section, suppose $h \geq 1$ is an integer.

Definition 1.3 The sequence of **corridor numbers** of height h and initial point $a \in \mathcal{C}^{(h)}$, denoted by $(c_{n,a}^{(h)})_{n \in \mathbb{N}_0}$, counts the total number of corridor paths of length n in the (h)-corridor, starting at $(0, a)$.

---

[2]A **walk** in a graph is defined as a finite sequence of adjacent vertices of the graph. Here, *adjacent* means *connected by an edge*. A **path graph** P_m is a graph on m vertices that are connected in a row by edges. A few helpful texts covering introductory graph theory include Brualdi [14], Epp [23], and Harris-Hirst-Mossinghoff [31], though we do not require an extensive understanding of graph theory for our purposes.

[3]I.e. the street along which he or she is walking is finite—this seems to me to be a much more reasonable scenario than the concept of the infinite street.

[4]An exhaustive bibliography of the work done in this field would be next to impossible to compile, but the interested reader may consult [7, 8, 12, 17, 18, 22, 25, 26, 33, 37, 38, 43, 45, 53, 56, 57].

[5]Our ideas are intimately related to the concept of *reflectable walks*, but more about this connection later.

Clearly, the number of corridor paths of length n is equal to the sum of the numbers of paths that end at (n, k) as k ranges in $C^{(h)}$. We describe the latter as **vertex numbers** and define them by way of a function of k.

Definition 1.4 Suppose $a \in C^{(h)}$ is fixed. Let $v_{n,a}^{(h)}(k)$ be the number of corridor paths in the (h)-corridor starting at $(0, a)$ and ending at (n, k), which for various k and n we call the **vertex numbers** for the corridor.

Example 1.1 Let $m \in \mathbb{N}$. The number of n-length walks in the path graph P_m starting at vertex a and ending at b $(1 \leq a, b \leq m)$ is given by the vertex number $v_{n,a}^{(m)}(b)$. The total number of n-length walks in P_m starting at vertex a is equal to the corridor number $c_{n,a}^{(m)}$.

In practice we omit the mention of the initial point a in the case that $a = 1$ (classic corridors); hence the notations $c_n^{(h)}$ and $v_n^{(h)}(k)$ are most often used. Furthermore, we may sometimes omit h if the height is fixed and clear by context, letting c_n and $v_n(k)$ denote the corridor and vertex numbers, respectively. The following formula relates Definitions 1.3 and 1.4.

$$c_n = \sum_{k=1}^{h} v_n(k) \qquad\qquad (1.1)$$

We refer to the function v_n as a **state function** for the corridor. At first, Definition 1.4 implies that the domain of v_n is the bounded discrete set $C^{(h)}$, however we will eventually extend the domain to all of $\mathbb{Z}$; indeed our methods will define state functions v_n over all of $\mathbb{C}$ but only those values on $C^{(h)}$ give the correct vertex numbers for a given corridor. The associated vector,

$$v_n = (v_n(1), v_n(2), \ldots, v_n(h)) \in \mathbb{C}^h,$$

may be called a **state vector**, generally using the notation v_n for either the vector or function. Note that $v_n \in \mathbb{Z}^h \subset \mathbb{C}^h$, and for various purposes, we may wish to emphasize either the integral or the complex nature of v_n.

The key point to be made here is that each state v_n can be found *recursively*. That is, after defining a suitable **initial state**, v_0, then certain operations may be applied to v_0 to produce $v_1, v_2, \ldots, v_n$ for any $n \in \mathbb{N}$. Our perspective is that each state v_n *describes* the possible corridor paths at step n, and *determines* those at step $n + 1$, as if it were some kind of physical system evolving with time, with each subsequent state being determined by predictable rules (just as physical laws may predict the motion of particles at future times given an initial configuration). With this interpretation, we view v_n as an element of a **state space** isomorphic to $\mathbb{C}^h$ (more about that point later, when we discuss the geometry of state vectors in Chap. 4). Figure 1.3 illustrates how a given initial state generates the next few states in a corridor.

At this point we recommend that the reader works out the progression of states and sequences of corridor numbers for some small h, say $h = 2, 3, 4$, and 5. Very likely you'll find yourself moving from one state to the next inductively by:

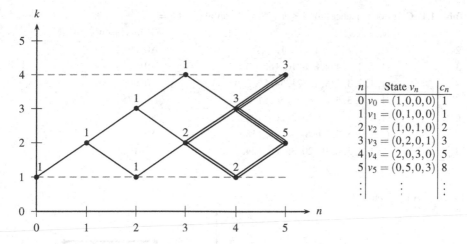

Fig. 1.3 The (4)-corridor, with all paths starting at $a = 1$. Each segment entering a particular vertex belongs to a distinct path. The number above each vertex is the value of $v_n(k)$ (values of 0 are not recorded on this diagram). For example, there are 5 paths of length 5 that end at vertex $(5, 2)$, three of which arrived from $(4, 3)$ and two from $(4, 1)$; hence, $v_5(2) = 5$

$$v_{n+1}(k) = \begin{cases} v_n(k-1) & \text{if } k = h, \\ v_n(k-1) + v_n(k+1) & \text{if } 1 < k < h, \\ v_n(k+1) & \text{if } k = 1 \end{cases} \qquad (1.2)$$

Note that when $h = 1$ (i.e. the corridor has height 1), there is only the trivial 0-length path. That is, $(c_n^{(1)})_{n \in \mathbb{N}_0} = (1, 0, 0, 0, \ldots)$. While this may not be a very exciting sequence, some other corridor sequences are rather interesting indeed. For example, as Fig. 1.3 suggests, the height $h = 4$ corridor sequence is exactly the sequence of *Fibonacci numbers*. Table 1.1 displays a few corridor sequences with initial point $a = 1$ along with their OEIS [54] labels (though most of these sequences are well-known for other reasons in the OEIS). The astute reader may notice that the sequences seem to stabilize as height increases, in the sense that for fixed n and a we have $c_{n,a}^{(h)} = c_{n,a}^{(h+1)} = c_{n,a}^{(h+2)} = \cdots$ for large enough h. We will discuss this further below as we explore *infinite height* corridors, but first let us discuss types of bounded lattice paths that appear in the combinatorial literature.

1.3 Dyck Paths and Motzkin Paths

Corridor paths as defined above are closely related to **Dyck paths** [26]. In the combinatorial literature, a Dyck path of order m is a lattice path from the origin to the point (m, m) that uses only up or right moves and does not cross the diagonal line $y = x$. By definition, an order m Dyck path has length $2m$. For our purposes, we will

Table 1.1 Corridor sequences for $2 \leq h \leq 8$, with initial point $a = 1$

h	$(c_n^{(h)})_{n \in \mathbb{N}_0}$	OEIS sequence
2	$(1, 1, 1, 1, 1, 1, 1, 1, 1, 1, \ldots)$	A000012
3	$(1, 1, 2, 2, 4, 4, 8, 8, 16, 16, \ldots)$	A016116
4	$(1, 1, 2, 3, 5, 8, 13, 21, 34, 55, \ldots)$	A000045
5	$(1, 1, 2, 3, 6, 9, 18, 27, 54, 81, \ldots)$	A038754
6	$(1, 1, 2, 3, 6, 10, 19, 33, 61, 108, \ldots)$	A028495
7	$(1, 1, 2, 3, 6, 10, 20, 34, 68, 116, \ldots)$	A030436
8	$(1, 1, 2, 3, 6, 10, 20, 35, 69, 124, \ldots)$	A061551

Fig. 1.4 A Dyck path of order 5

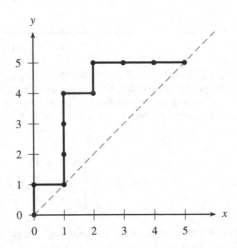

assume the path is drawn above the line $y = x$, as in Fig. 1.4. It is well known that the number of Dyck paths of order $m \geq 0$ is equal to the mth **Catalan number** [14, 15], $C_m = \frac{1}{m+1}\binom{2m}{m}$. We are interested in the following variation of Dyck paths found in Krattenthaler-Mohanty [37] and elsewhere.

Definition 1.5 Let $s, t \in \mathbb{Z}$ such that $t \geq 0 \geq s$ and $p, q \in \mathbb{Z}$ such that $p + t \geq q \geq p + s$. A **restricted** Dyck path is lattice path from the origin to (p, q) that uses only up or right moves and does not cross either of the lines $y = x + s$ or $y = x + t$. The number of such paths is denoted by $D(p, q; s, t)$.

Krattenthaler and Mohanty state two related formulae for D, one involving trigonometric functions ([37], Eq. (5)), which we will recover in this text by Fourier methods (see Theorem 2.1), and another ([37], Eq. (9)) involving only the binomial coefficients,

$$D(p, q; s, t) = \sum_{k \in \mathbb{Z}} \left[\binom{p+q}{p - k(t - s + 2)} - \binom{p+q}{p - k(t - s + 2) + t + 1} \right],$$

(1.3)

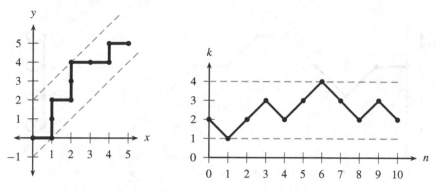

Fig. 1.5 *(Left)* A restricted Dyck path of length 10 ending at (5, 5), between $y = x - 1$ and $y = x + 2$ ($s = -1$ and $t = 2$). *(Right)* The corresponding path in the (4)-corridor, with initial point (0, 2)

where we follow the convention that $\binom{n}{k} = 0$ if $k < 0$ or $k > n$ (see Eq. (1.6)). There is an affine transformation taking restricted Dyck paths of Definition 1.5 into corridor paths. With p, q, s, t as in Definition 1.5, map the point $(p, q) \mapsto (p + q, q - p - s + 1)$. Then the lines $y = x + s$ and $y = x + t$ map to the bottom and top boundaries of the corridor, $y = 1$ and $y = t - s + 1$, respectively. The origin maps to $(0, -s + 1)$. The result is the (h)-corridor, where $h = t - s + 1$, with initial point $a = -s + 1$ (see Fig. 1.5, for example), hence we have a direct connection to our vertex numbers:

$$D(p, q, s, t) = v_{p+q, \ s|1}^{(t-s+1)}(q - p - s + 1) \tag{1.4}$$

A **Motzkin path** is by definition a lattice path in $\mathbb{N}_0 \times \mathbb{N}_0$ starting at the origin, ending on the horizontal axis, and such that each step in the path is either up-right, right, or down-right. We will be interested in these kinds of **three-way** paths but within the (h)-corridor, $\mathbb{N}_0 \times \mathcal{C}^{(h)}$, beginning and ending at arbitrary heights, as in Fig. 1.6. Such paths may be interpreted as walks in the graph P_h' obtained from P_h by adding a loop to each vertex [25].[6]

If $m_n(k) = m_{n,a}^{(h)}(k)$ denotes the number of three-way paths beginning at $(0, a)$ and ending at (n, k) in the (h)-corridor, then an inductive formula similar to (1.2) computes the values $n \geq 0$, with $m_0(k)$ defined appropriately to indicate the initial point of the paths.

$$m_{n+1}(k) = \begin{cases} m_n(k - 1) + m_n(k) & \text{if } k = h, \\ m_n(k - 1) + m_n(k) + m_n(k + 1) & \text{if } 1 < k < h, \\ m_n(k) + m_n(k + 1) & \text{if } k = 1 \end{cases} \tag{1.5}$$

---

[6]Alternatively, one could imagine a *drunkard's walk* in which the drunkard frequently stops to think about which way to move next.

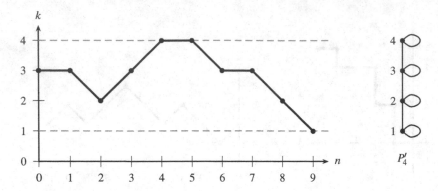

Fig. 1.6 A three-way corridor path of length 9 in the (4)-corridor with initial point $a = 3$. The corresponding walk in P_4' visits the vertices in the order, 3, 3, 2, 3, 4, 4, 3, 3, 2, 1, where a repeated vertex implies that the walk traverses a loop

1.4   The Binomial Coefficients and Unbounded Lattices

Throughout this text we will encounter the **binomial coefficients**. We assume the reader is familiar with their definition, but for completeness, we include an inductive definition. Assume $n \in \mathbb{N}_0$ and $k \in \mathbb{Z}$.

$$\binom{0}{k} = \begin{cases} 1 & \text{if } k = 0, \\ 0 & \text{if } k \neq 0, \end{cases} \quad \text{and for } n \geq 0, \quad \binom{n+1}{k} = \binom{n}{k-1} + \binom{n}{k} \quad (1.6)$$

Note that this definition implies $\binom{n}{k} = 0$ when $k < 0$ or $k > n$, and $\binom{n}{k} = 1$ whenever $k = 0$ or $k = n$. We also take the convention that $\binom{n}{k} = 0$ if $k \notin \mathbb{Z}$. Formula (1.6) generates an array of numbers called **Pascal's triangle**, which we will represent as a doubly indexed array.

Definition 1.6 The **Pascal array** is the array whose row n, column k entry is equal to $\binom{n}{k}$ where $n \in \mathbb{N}$, $k \in \mathbb{Z}$.

$n \setminus k$	...	−1	0	1	2	3	4	5	...
0	...	0	1	0	0	0	0	0	...
1	...	0	1	1	0	0	0	0	...
2	...	0	1	2	1	0	0	0	...
3	...	0	1	3	3	1	0	0	...
4	...	0	1	4	6	4	1	0	...
5	...	0	1	5	10	10	5	1	...
⋮		⋮	⋮	⋮	⋮		⋮	⋮	

The binomial coefficients are extremely useful in mathematics; for example, we have the Binomial Theorem, which serves to rewrite a power of a sum as a sum of powers.

Theorem 1.1 (Binomial Theorem) *For $n \geq 0$, and for all $a, b \in \mathbb{C}$,*

$$(a + b)^n = \sum_{k=0}^{n} \binom{n}{k} a^{n-k} b^k, \qquad (1.7)$$

where we interpret $0^0 = 1$.

As an application to lattice path enumeration, consider paths in $\mathbb{N}_0 \times \mathbb{N}_0$ starting at the origin with no upper bound on the lattice, such that each step in the path is either right or up-right (see Fig. 1.7). Let $u_n(k)$ be the number of such paths ending at (n, k). Note that there is a recursive formula for $u_n(k)$:

$$u_0(k) = \begin{cases} 1 & \text{if } k = 0, \\ 0 & \text{if } k \neq 0, \end{cases} \quad \text{and for } n \geq 0, \ u_{n+1}(k) = u_n(k-1) + u_n(k) \qquad (1.8)$$

Of course, Formula (1.8) is nothing more than Formula (1.6) but with a different label for the function, proving that $u_n(k) = \binom{n}{k}$. Furthermore, it is easy to find the sequence of corridor numbers in this case,

$$c_n = \sum_{k=0}^{n} u_n(k) = \sum_{k=0}^{n} \binom{n}{k} = 2^n,$$

corresponding to the two choices to move at each step.

Fig. 1.7 The number of lattice paths in $\mathbb{N}_0 \times \mathbb{N}_0$ whose moves are either right or up-right, starting at the origin and ending at (n, k), is equal to the binomial coefficient $\binom{n}{k}$

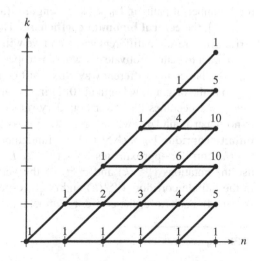

Remark 1.2 For those who may be familiar with the techniques of **generating functions** [14, 26, 64], Eq. (1.7) provides a generating function for the nth row of Pascal's array:

$$(1 + x)^n = \sum_{k \geq 0} \binom{n}{k} x^k = \binom{n}{0} + \binom{n}{1} x + \binom{n}{2} x^2 + \binom{n}{3} x^3 + \cdots + \binom{n}{n} x^n$$

If "1" represents a right move and "x" represents an up-right move (as in Fig. 1.7, then the number of paths ending at (n, k) is exactly the coefficient of x^k in the generating function $g(x) = (1 + x)^n$, which is precisely $\binom{n}{k}$ by the Binomial Theorem. Then the total count of paths of length n is found by summing the coefficients, or equivalently, substituting $x = 1$ into the generating function to get $c_n = g(1) = (1 + 1)^n = 2^n$.

By a simple transformation, we can count the number of paths in $\mathbb{N}_0 \times \mathbb{Z}$ starting at the origin, in which the allowable moves are up-right and down-right, as illustrated in Fig. 1.8. That is, we are counting the number of two-way paths in a totally unrestricted "corridor" (though corridors without "walls" are hardly corridors at all). Let $w_n(k)$ be the number of paths ending at $(n, k) \in \mathbb{N}_0 \times \mathbb{Z}$. Then we have the following formula.

$$w_n(k) = \binom{n}{\frac{1}{2}(n + k)} \tag{1.9}$$

Next, let's consider corridor paths in $\mathbb{N}_0 \times \mathbb{N}_0$. This time, there is a lower boundary (the x-axis), but still no upper boundary. This "corridor" can be used to address Bertrand's Ballot Problem [1, 10, 61]. In particular we may ask, *How many of the 2^n paths of length n in the $\mathbb{N}_0 \times \mathbb{Z}$ corridor are in fact paths in the $\mathbb{N}_0 \times \mathbb{N}_0$ corridor?*

It is well known that the total number of paths of length $n = 2m$ in $\mathbb{N}_0 \times \mathbb{N}_0$ is given by $\binom{2m}{m}$, but there is an analogous result in the cases in which n is odd. For any number $x \in \mathbb{R}$, define the **floor** $\lfloor x \rfloor$ of x to be the integer k such that $k \leq x < k + 1$. In other words, $\lfloor x \rfloor$ rounds x to the nearest integer below (or equal) to x. Then the total number of paths in $\mathbb{N}_0 \times \mathbb{N}_0$ of length n (and beginning at the origin) is equal to $\binom{n}{\lfloor n/2 \rfloor}$, the **central binomial coefficient**.[7] However we will demonstrate how to arrive at this result in different ways, and we will be able to generalize it significantly.

In order to better motivate our work in proper corridors $\mathbb{N}_0 \times \mathcal{C}^{(h)}$, we interpret the $\mathbb{N}_0 \times \mathbb{N}_0$ lattice in a different way. First shift up by one unit so that all paths that had begun at the origin now begin at $(0, 1)$ in the lattice $\mathbb{N}_0 \times \mathbb{N}$—that is, the horizontal line $y = 1$ becomes the lower boundary rather than the x-axis. Even though there is no upper boundary, we still refer to this as a corridor, or more specifically, the **infinite corridor** $\mathbb{N}_0 \times \mathcal{C}^{(\infty)}$ (whose fundamental region is $\mathcal{C}^{(\infty)} = \mathbb{N}$). Note that the (h)-corridor approximates $\mathbb{N}_0 \times \mathcal{C}^{(\infty)}$ as h increases, and so it seems natural to use the notations $v_n^{(\infty)}(k)$ and $c_n^{(\infty)}$ for the vertex numbers and corridor numbers in the infinite corridor ($v_{n,a}^{(\infty)}(k)$ and $c_{n,a}^{(\infty)}$, respectively, if the initial point $a \geq 1$ is specified). As before, the corridor numbers are given by a sum of vertex numbers,

---

[7]OEIS sequence A001405.

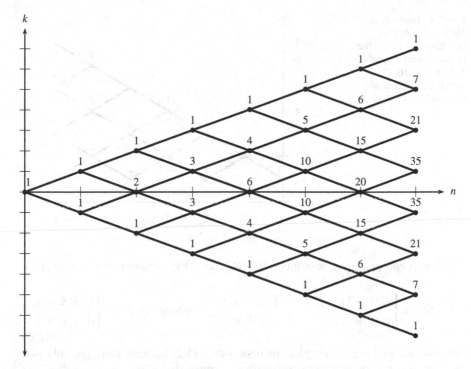

Fig. 1.8 Unrestricted lattice paths in $\mathbb{N}_0 \times \mathbb{Z}$ beginning at the origin

Table 1.2 States and corridor numbers for the infinite corridor

n	State v_n	c_n
0	$v_0 = (1, 0, 0, \ldots)$	1
1	$v_1 = (0, 1, 0, 0, \ldots)$	1
2	$v_2 = (1, 0, 1, 0, 0, \ldots)$	2
3	$v_3 = (0, 2, 0, 1, 0, 0, \ldots)$	3
4	$v_4 = (2, 0, 3, 0, 1, 0, 0, \ldots)$	6
5	$v_5 = (0, 5, 0, 4, 0, 1, 0, 0, \ldots)$	10
6	$v_6 = (5, 0, 9, 0, 5, 0, 1, 0, 0, \ldots)$	20
7	$v_7 = (0, 14, 0, 14, 0, 6, 0, 1, 0, 0, \ldots)$	35
8	$v_8 = (14, 0, 28, 0, 20, 0, 7, 0, 1, 0, 0, \ldots)$	70
9	$v_9 = (0, 42, 0, 48, 0, 27, 0, 8, 0, 1, 0, 0, \ldots)$	126
$\vdots$	$\vdots$	$\vdots$

$c_n^{(\infty)} = \sum_{k=0}^{\infty} v_n^{(\infty)}(k)$. Here, even though the sum is over $k \in \mathbb{N}$, there are only finitely many nonzero terms for any given $n \in \mathbb{N}_0$. Table 1.2 contains state vectors (vertex numbers) and corridor numbers for small values of n (see Fig. 1.9).

Fig. 1.9 Two-way paths in $\mathbb{N}_0 \times \mathcal{C}^{(\infty)}$. Adding vertically, we obtain the infinite corridor numbers, 1, 1, 2, 3, 6, 10, 20, ..., which are the central binomial coefficients, $\binom{n}{\lfloor n/2 \rfloor}$

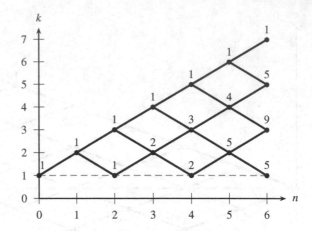

We may compute the state numbers $v_n^{(\infty)}(k)$ inductively in a similar way as in (1.2).

$$v_{n+1}(k) = \begin{cases} v_n(k-1) + v_n(k+1) & \text{if } k > 1, \\ v_n(k+1) & \text{if } k = 1, \end{cases} \quad \text{where} \quad v_0(k) = \begin{cases} 1 & \text{if } k = 1, \\ 0 & \text{if } k > 1 \end{cases}$$

(1.10)

Observe that (1.10) is a "simpler" recursion than (1.2) because there are only two cases rather than three. Next we describe a "trick" that allows one to reduce the number of cases to one. We extend the domain of $v_n(k)$ to all $k \in \mathbb{Z}$ in such a way so that each v_n is *antisymmetric* with respect to the origin—in other words, each v_n is an *odd* function.[8] In particular, the initial state v_0 is defined by:

$$v_0(k) = \begin{cases} 1 & \text{if } k = 1, \\ -1 & \text{if } k = -1, \\ 0 & \text{if } k \neq \pm 1 \end{cases}$$

(1.11)

The effect is to introduce paths beginning at $(0, -1)$, but with *negative weight*, which serve to cancel away paths from $(0, 1)$ that descend below height 1. We say that the extended lattice $\mathbb{N}_0 \times \mathbb{Z}$ together with the antisymmetry of the state functions v_n constitute a **dual corridor structure** (see Fig. 1.10). For added clarity, when our state vectors are sequences indexed by $\mathbb{Z}$, we often indicate the $k = 0$ term by writing it in bold face. The states corresponding to Fig. 1.10 are:

---

[8]This type of extension by antisymmetry, prevalent throughout the signal processing literature, should not really be considered a *trick*, but rather a useful method to aid in analyzing certain kinds of sequences (see e.g., [21]).

Fig. 1.10 The dual corridor structure for the infinite corridor. Paths beginning at $(0, 1)$ have weight 1, and those beginning at $(0, -1)$ have weight -1. The net effect is that path counts cancel at $k = 0$

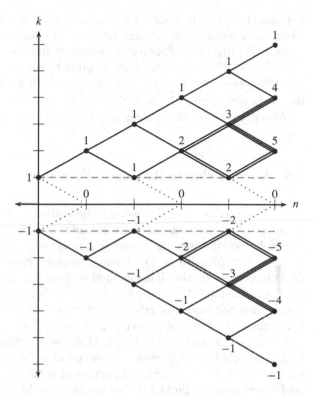

$$
\begin{aligned}
v_0 &= (\ldots, \quad 0, \quad 0, \quad 0, \quad 0, \quad 0, -1, \textbf{0}, \ 1, \ 0, \ 0, \ 0, \ 0, \ 0, \ldots) \\
v_1 &= (\ldots, \quad 0, \quad 0, \quad 0, \quad 0, -1, \quad 0, \textbf{0}, \ 0, \ 1, \ 0, \ 0, \ 0, \ 0, \ldots) \\
v_2 &= (\ldots, \quad 0, \quad 0, \quad 0, -1, \quad 0, -1, \textbf{0}, \ 1, \ 0, \ 1, \ 0, \ 0, \ 0, \ldots) \\
v_3 &= (\ldots, \quad 0, \quad 0, -1, \quad 0, -2, \quad 0, \textbf{0}, \ 0, \ 2, \ 0, \ 1, \ 0, \ 0, \ldots) \\
v_4 &= (\ldots, \quad 0, -1, \quad 0, -3, \quad 0, -2, \textbf{0}, \ 2, \ 0, \ 3, \ 0, \ 1, \ 0, \ldots) \\
v_5 &= (\ldots, -1, \quad 0, -4, \quad 0, -5, \quad 0, \textbf{0}, \ 0, \ 5, \ 0, \ 4, \ 0, \ 1, \ldots)
\end{aligned}
$$

In order to count the corridor paths, observe that the recursive formula for v_n now takes the form,

$$v_{n+1}(k) = v_n(k - 1) + v_n(k + 1), \tag{1.12}$$

where v_0 is defined as in Eq. (1.11). An explicit formula for $v_n(k)$ is now quite straightforward to put together. Recall $w_n(k)$ from Eq. (1.9). The *seed* value $v_0(1) = +1$ contributes $(+1)w_n(k - 1)$ to $v_n(k)$, and the seed value $v_0(-1) = -1$ contributes $(-1)w_n(k + 1)$. Therefore, we obtain the following result.

$$v_n^{(\infty)}(k) = \binom{n}{\frac{1}{2}(n + k - 1)} - \binom{n}{\frac{1}{2}(n + k + 1)} \tag{1.13}$$

Formula (1.13) may be regarded as a *superposition* of paths, those starting at $(0, 1)$ contributing positively to the total, and those starting at $(0, -1)$ contributing negatively, with perfect cancellation at the boundary line $k = 0$.

The $\mathbb{N}_0 \times \mathcal{C}^{(\infty)}$ corridor numbers are given by summing $v_n^{(\infty)}(k)$ over $k \in \mathbb{N}$. The sum telescopes, leaving only $\binom{n}{\frac{1}{2}(n-1)} + \binom{n}{\frac{1}{2}n}$, which is equivalent to $c_n^{(\infty)} = \binom{n}{\lfloor n/2 \rfloor}$, that is, the central binomial coefficients as expected.

We will have more to say about infinite corridors in Sect. 2.5.

1.5 Relationship to Reflection Principle

The technique of introducing negative weight paths to cancel paths that strayed below the boundary of the corridor may be regarded as an encoding of André's [3] **reflection principle**.[9] We illustrate the connection using our $\mathbb{N}_0 \times \mathcal{C}^{(\infty)}$ corridor. First count all paths from $(0, 1)$ to (n, k) without restriction. Now if a path p from $(0, 1)$ to (n, k) hits the x-axis, then it has crossed the boundary line $k = 1$ and must be taken away from the total. This is done in the following manner: consider the **conjugate** path p^* obtained from p by reflecting all points across the line $k = 0$ up to the first intersection with $k = 0$, and leaving the remaining points unreflected. Note that p^* begins at $(-1, 0)$ and ends at (n, k). More importantly, the set of all paths from $(-1, 0)$ to (n, k) is in bijection with the set of paths from $(0, 1)$ to (n, k) that hit the line $k = 0$. Thus, the required path count is the simple difference of binomial coefficients given by Eq. (1.13). The conjugate paths play the role of our negative weighted paths in the dual corridor. See Fig. 1.11 for a graphical interpretation.

The corridor vertex counts $v_n^{(h)}(k)$ may also be obtained by the reflection principle. This time inclusion-exclusion must be used to obtain the correct counts. This corresponds to repeatedly reflecting the corridor $\mathbb{N}_0 \times \mathcal{C}^{(h)}$ above and below alternating the signs (weights) of the vertex numbers each time so that cancellation occurs at $k = 0$ and $k = h + 1$. Indeed, $v_n(k) = 0$ whenever k is a multiple of $h + 1$. Compare Fig. 1.12 with the corridor pictured in Fig. 1.3.

This observation led the authors [6] to consider defining the vertex numbers $v_n^{(h)}(k)$ for all $k \in \mathbb{Z}$ so that they are antisymmetric in k and periodic with period $2d$, where $d = h + 1$. Finally, once we have defined periodic functions, it is a natural next step to consider the discrete Fourier transform. Our methods, which we will define precisely in the following chapters, serve as an alternative way of encoding the reflection principle and inclusion-exclusion using only elementary techniques from analysis.

---

[9]More recent exposition of the reflection principle is surveyed in [1, 17, 30, 37].

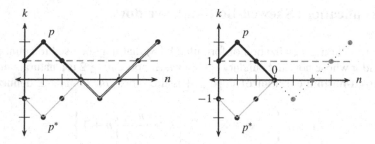

Fig. 1.11 *(Left)* Path p (in black) which hits the horizontal line $k = 0$, and its conjugate p^* (in light gray). *(Right)* Interpreting p^* as having a negative weight, the paths p and p^* cancel

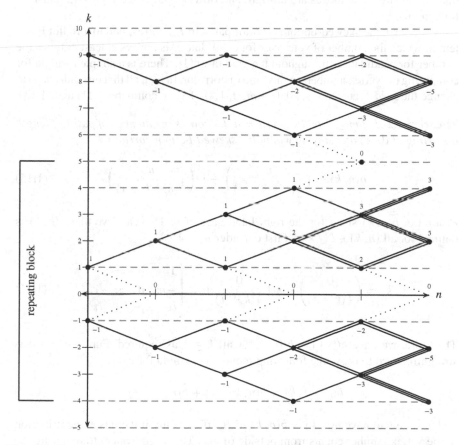

Fig. 1.12 The dual corridor structure for $h = 4$ and $a = 1$ with nonzero vertex numbers shown. This picture extends to all of $\mathbb{N}_0 \times \mathbb{Z}$, repeating vertically in blocks of size $2(h + 1) = 10$

1.6   Application: Skewed-Bottom Corridors

We now present a generalization of formula (1.13) that may be used to count paths
in a corridor whose bottom boundary is *skewed*. Suppose $m \geq 2$ is a natural number.
The **skewed-bottom corridor** of slope $\frac{m-1}{m+1}$ is the subset $B_m \subseteq \mathbb{N}_0 \times \mathbb{Z}$ defined by:

$$B_m = \left\{ (n, k) \in \mathbb{N}_0 \times \mathbb{Z} \; \middle| \; k \geq \frac{m-1}{m+1} n + 1 \right\} \tag{1.14}$$

A corridor path in the skewed-bottom corridor can be defined as in Definition 1.2,
that is, the allowable moves are up-right and down-right.[10] See Figs. 1.13 and 1.14
for examples.

As defined, the skewed-bottom corridor paths in B_m correspond to a Ballot Problem in which the number of votes cast for candidate A is always at least m times the
number for candidate B throughout the vote tally [1]. There is a simple formula for
computing the vertex numbers in this kind of corridor. In fact, all that one must do is to
change the *seed* vector to $(\ldots, 0, 0, -m, \mathbf{0}, 1, 0, 0, \ldots)$; compare (1.11) and (1.13).

Theorem 1.2 *Let $m \geq 2$ and suppose $b(n, k)$ counts the number of paths of length
n starting at $(0, 1)$ and staying within the skewed-bottom corridor B_m.*

$$b(n, k) = \binom{n}{\frac{1}{2}(n + k - 1)} - m \binom{n}{\frac{1}{2}(n + k + 1)} \tag{1.15}$$

Proof Let $\widetilde{b}(n, k)$ stand for the right hand side of (1.15). Observe that $\widetilde{b}(n, k)$ is
defined for all $(n, k) \in \mathbb{N}_0 \times \mathbb{Z}$. First consider $n = 0$.

$$\widetilde{b}(0, k) = \binom{0}{\frac{1}{2}(k - 1)} - m \binom{0}{\frac{1}{2}(k + 1)} = \begin{cases} 1 & \text{if } k = 1, \\ -m & \text{if } k = -1, \\ 0 & \text{if } k \neq \pm 1 \end{cases} \tag{1.16}$$

Therefore, we have $\widetilde{b}(0, k) = b(0, k)$ for all $k \geq 1$ as required. Furthermore, it is
straightforward to prove the following recursive formula for $n \geq 0$.

$$\widetilde{b}(n + 1, k) = \widetilde{b}(n, k - 1) + \widetilde{b}(n, k + 1) \tag{1.17}$$

Now we will have $\widetilde{b}(n, k) = b(n, k)$ in B_m if and only if there is no contribution
to the vertex number counts from outside of B_m. This is equivalent to requiring the
value of $\widetilde{b}$ to be 0 at any vertex (n, k) that connects to a vertex of B_m from below
(see Fig. 1.13). Note that the nonzero vertex numbers in the corridor B_m occur on the
boundary line only at points of the form:

---

[10]However it would be interesting to consider three-way moves as well.

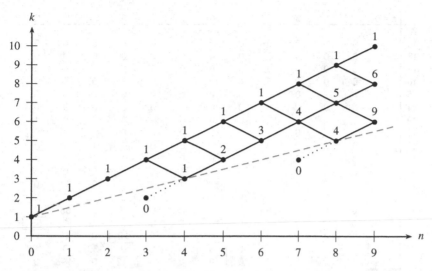

Fig. 1.13 The skewed-bottom corridor B_3. The bottom boundary is the line $k = \frac{3-1}{3+1}n + 1 = \frac{n}{2} + 1$. The numbers at each vertex count the number of distinct paths from $(0, 1)$ to that vertex using only up-right and down-right moves and staying within B_3. Zero vertex numbers are shown at key points to illustrate the proof of Theorem 1.2

$$\begin{cases} n = (m + 1)t, \\ k = (m - 1)t + 1, \end{cases} \quad t \in \mathbb{N}$$

Therefore, we require that the vertex numbers at points that are down and left from these points must be 0. In particular, we require

$$\widetilde{b}((m + 1)t - 1, (m - 1)t) = 0 \tag{1.18}$$

for all $t \in \mathbb{N}$. Equation (1.18) can easily be established from the familiar factorial formula for the binomial coefficients:

$$\binom{n}{k} = \frac{n!}{k!(n - k)!}$$

$\square$

Exercises

1.1 Construct a diagram and table similar to those in Fig. 1.3 to compute the vertex and corridor numbers for the (5)-corridor, up to $n = 12$.

1.2 Find the vertex numbers $v_{n,2}^{(4)}(k)$ for $(n, k) \in \{0, 1, \ldots, 10\} \times \mathcal{C}^{(4)}$. Do the same for $v_{n,3}^{(4)}(k)$ and $v_{n,4}^{(4)}(k)$. What similarities do you notice?

k	n = 0	1	2	3	4	5	6	7	8	9	10	11	12
20													
19		0	0	0	0	0	0	0	0	0	0	0	0
18		0	0	0	0	0	0	0	0	0	0	0	0
17		0	0	0	0	0	0	0	0	0	0	0	0
16		0	0	0	0	0	0	0	0	0	0	0	0
15		0	0	0	0	0	0	0	0	0	0	0	0
14		0	0	0	0	0	0	0	0	0	0	0	0
13		0	0	0	0	0	0	0	0	0	0	0	1
12		0	0	0	0	0	0	0	0	0	0	1	0
11		0	0	0	0	0	0	0	0	0	1	0	10
10		0	0	0	0	0	0	0	0	1	0	9	0
9		0	0	0	0	0	0	0	1	0	8	0	42
8		0	0	0	0	0	0	1	0	7	0	33	0
7		0	0	0	0	0	1	0	6	0	25	0	88
6		0	0	0	0	1	0	5	0	18	0	55	0
5		0	0	0	1	0	4	0	12	0	30	0	55
4		0	0	1	0	3	0	7	0	12	0	0	0
3		0	1	0	2	0	3	0	0	0	-30	0	-198
2		1	0	1	0	0	0	-7	0	-42	0	-198	0
1	1	0	0	0	-2	0	-10	0	-42	0	-168	0	-660
0		-1	0	-3	0	-10	0	-35	0	-126	0	-462	0
-1	-2	0	-3	0	-8	0	-25	0	-84	0	-294	0	-1056
-2		-2	0	-5	0	-15	0	-49	0	-168	0	-594	0
-3		0	-2	0	-7	0	-24	0	-84	0	-300	0	-1089
-4		0	0	-2	0	-9	0	-35	0	-132	0	-495	0
-5		0	0	0	-2	0	-11	0	-48	0	-195	0	-770
-6		0	0	0	0	-2	0	-13	0	-63	0	-275	0
-7		0	0	0	0	0	-2	0	-15	0	-80	0	-374
-8		0	0	0	0	0	0	-2	0	-17	0	-99	0
-9		0	0	0	0	0	0	0	-2	0	-19	0	-120
-10		0	0	0	0	0	0	0	0	-2	0	-21	0
-11		0	0	0	0	0	0	0	0	0	-2	0	-23
-12		0	0	0	0	0	0	0	0	0	0	-2	0
-13		0	0	0	0	0	0	0	0	0	0	0	-2

Fig. 1.14 Using a spreadsheet program to calculated corridor numbers. Entries correspond to the values of $v_n(k)$ in B_2. The seed v_0 defined by $v_0(1) = 1$, $v_0(-1) = -2$, and $v_0(x) = 0$ for $x \neq \pm 1$ produces a skewed-bottom corridor

1.3 Suppose $1 \leq a, k \leq h$. Prove that $v_{n,a}^{(h)}(k) = v_{n,h+1-a}^{(h)}(h + 1 - k)$, and use this result to show that $c_{n,a}^{(h)} = c_{n,h+1-a}^{(h)}$.

1.4 Prove that $\binom{n}{k} = \frac{n!}{k!(n-k)!}$, for $n, k \in \mathbb{N}_0$, $0 \leq k \leq n$, using the recursive definition (1.6).

1.5 By grouping two terms at a time and using the Binomial Theorem twice, expand $(a + b + c)^n$ into a sum of terms. Use your formula to find the coefficient of x^5 in $(1 + x + x^2)^{10}$.

1.6 *(Requires generating functions)* Suppose $g(x)$ is a generating function for the constant series $(1, 1, 1, 1, \ldots)$. Using a shift (i.e. $xg(x)$), derive a closed formula for $g(x)$. Explain how this function can be used to find the sum of a **geometric series**, $1 + x + x^2 + x^3 + \cdots$ when $|x| < 1$.

1.7 For $n \geq 1$, define the nth Catalan number C_n as follows (see e.g. [14, 57]): On the square $\{0, 1, \ldots, n\} \times \{0, 1, \ldots, n\}$, C_n counts the number of paths from $(0, 0)$ to (n, n), where the path moves up a unit or right a unit, and the path never falls below the diagonal (see Fig. 1.15). Restate the definition of C_n in terms of unbounded corridor paths and then use Formula (1.13) to show that $C_n = \frac{1}{n+1}\binom{2n}{n}$.

1.8 A **Catalan tree** is a data structure in computer science which corresponds to a hierarchical structure such as the open and close parentheses of an expression. The *height* of a Catalan tree is the number of nodes on a maximal path from root to leaf (see Fig. 1.16), and its *size* is the total number of edges. There is a bijection between Catalan trees of height q and size m with corridor paths found by making a *pre-order traversal* [17]:

> We imagine that the tree is a roadmap and our avatar plans a tour starting at the root as follows: We take the rightmost unvisited road (from the avatar's viewpoint), else we backtrack...

Let the root node correspond to the point $(0, 1)$ in the lattice. Each downward (forward) move on the tree corresponds to an up-right move in the corridor, while an upward (backtrack) move on the tree corresponds to down-right move in the corridor. Note, the total number of upward moves can never exceed the number of downward moves, hence the corresponding corridor path is restricted to $\mathbb{N}_0 \times \mathbb{N}$. Using your table from Exercise 1.1, count the number of distinct Catalan trees of height at most 5 and size 6.

1.9 Find the affine transformation inverse to $(p, q) \mapsto (p + q, q - p - s + 1)$ and use it to write $v_{n,a}^{(h)}(x)$ in terms of the function D from Definition 1.5 and Eq. (1.4). Then use Eq. (1.3) to write a formula for $v_{n,a}^{(h)}(x)$ as a sum of binomial coefficients.

1.10 By drawing a diagram similar to Fig. 1.3, and/or by using formula (1.5), find the number of three-way paths in the (3)-corridor, starting at $a = 1$, of length $n = 0, 1, 2, 3, \ldots$ (as far as you want to go). Check your work by looking up OEIS sequence A000129, also known as the sequence of *Pell numbers*.

Fig. 1.15 A path contributing to C_3

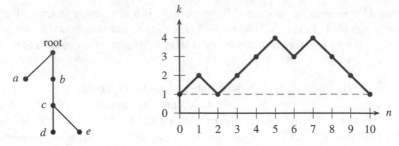

Fig. 1.16 *(Left)* A Catalan tree of height 4 and size 5, with pre-order traversal: root-a-root-b-c-d-c-e-c-b-root. *(Right)* The corresponding corridor path

1.11 Let $r \geq 2$, and suppose $n, k_1, k_2, \ldots, k_r \in \mathbb{N}_0$. Define the **multinomial coefficient** as follows.

$$\binom{n}{k_1, k_2, \ldots, k_r} = \frac{n!}{k_1! k_2! \cdots k_r!}$$

There is a result called the **Multinomial Theorem** analogous to Theorem 1.1,

$$(a_1 + a_2 + \cdots + a_r)^n = \sum_{\substack{k_j \geq 0, \\ k_1 + \cdots + k_r = n}} \binom{n}{k_1, k_2, \ldots, k_r} a_1^{k_1} a_2^{k_2} \cdots a_r^{k_r}, \qquad (1.19)$$

for all $a_1, a_2, \ldots, a_r \in \mathbb{C}$.

(a) Use Formula (1.19) to write $(a + b + c)^5$ as a sum of terms (compare Exercise 1.5).

(b) Consider lattice paths in $\mathbb{N}_0 \times \mathbb{N}_0$ in which there are three valid moves: right $(1, 0)$, up-right $(1, 1)$, and up-up-right $(1, 2)$. Use a generating function argument to prove that the number of such paths of length n is equal to the following sum.

$$\sum_{p+q+r=k} \binom{n}{p, q, r}$$

What if the available moves are $(1, 0)$, $(1, 1)$, ..., $(1, q)$?

1.12 (Note: This problem requires familiarity with **generating functions**.) Consider the family of *Laurent polynomials*[11] defined recursively by $p_0(x) = x - x^{-1}$,

---

[11]A **Laurent polynomial** is a function of the form $g(x) = \sum_{k=M}^{N} a_k x^k$, where $M, N \in \mathbb{Z}$. In particular, negative powers are permitted.

and for $n \geq 0$, $p_{n+1}(x) = (x + x^{-1})p_n(x)$. For $(n, k) \in \mathbb{N}_0 \times \mathbb{Z}$, let $p_{n,k}$ be the coefficient of x^k in $p_n(x)$. Show that $(p_{0,k})_{k \in \mathbb{Z}} = (\ldots, 0, 0, -1, \mathbf{0}, 1, \ldots, 0, 0, \ldots)$ and that for $n \geq 0$, we have $p_{n+1,k} = p_{n,k-1} + p_{n,k+1}$. Thus, $p_n(x)$ is a generating function for $(v_n^{(\infty)}(k))_{k \in \mathbb{Z}}$. Use the Binomial Theorem to expand $p_n(x)$, arriving at Eq. (1.13) by different means.

1.13 Establish Eq. (1.17). Then prove (1.18) using (1.15).

1.14 Consider two-way lattice paths beginning at the origin and never going below the line $y = x/r$. Use Theorem 1.2 to verify that the number of such paths that return to the line $y = x/3$ after $3k$ steps is given by the formula,[12] $\frac{1}{2k+1}\binom{3k}{k}$.

1.15 Use a spreadsheet program (e.g. OpenOffice Calc[13]) to compute $v_n^{(\infty)}(k)$ for a range of (n, k) values. The first column should contain the entries of v_0 (the *seed* values defined by (1.11)), and each successive column should be able to calculate the entries of v_1, v_2, v_3, etc. based the simple recursive formula (1.12).[14] Then use the seed values defined by (1.16) for various choices of m. Compare your results for $m = 2$ with those in Fig. 1.14.

Research Questions

1.16 Define a **skewed-top corridor**[15] as suggested by Fig. 1.17. Explore the vertex numbers and corridor numbers for various slopes and initial gaps. What patterns exist among the "corner" numbers where the corridor meets the upper boundary line?

1.17 Explore corridors bounded by arbitrary lines in $\mathbb{N}_0 \times \mathbb{N}_0$. For example, what can be said if both the top and the bottom boundary lines are skewed?

1.18 Research the **Steck matrix** [36, 42, 58, 59]. Steck matrices can be used to count lattice paths in regions of the plane bounded by arbitrary upper and lower boundaries using determinants. Set up a Steck matrix that would compute the vertex numbers in the (4)-corridor, thereby showing that the Fibonacci numbers, F_n for $n \geq 2$, may be found by determinants of matrices in the following way.

---

[12]See also OEIS sequence A001764.

[13]https://www.openoffice.org/product/calc.html.

[14]The reason that a spreadsheet program is especially useful is that the recursive formula only needs to be defined once. Then you can "drag" that formula up, down, and to the right to fill out the spreadsheet.

[15]See also [9].

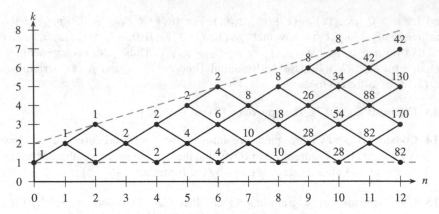

Fig. 1.17 A skewed-top corridor with top boundary slope $\frac{1}{2}$ and initial gap 1

$$
F_{2m} = \begin{vmatrix} 2 & 1 & 0 & 0 & \cdots & 0 \\ 1 & 3 & 1 & 0 & & \vdots \\ 0 & 1 & 3 & \ddots & & \\ \vdots & & \ddots & \ddots & & 1 \\ 0 & \cdots & & & 1 & 3 \end{vmatrix} \quad \text{if } m \geq 1, \quad F_{2m+1} = \begin{vmatrix} 3 & 1 & 0 & 0 & \cdots & 0 \\ 1 & 3 & 1 & 0 & & \vdots \\ 0 & 1 & 3 & \ddots & & \\ \vdots & & \ddots & \ddots & & 1 \\ 0 & \cdots & & & 1 & 3 \end{vmatrix} \quad \text{if } m \geq 1
$$

Explore other Steck determinants and their relationships to corridor numbers or their generalizations (e.g. skewed-top or skewed-bottom corridors).

Chapter 2
One-Dimensional Lattice Walks

In this chapter we show how the Fourier transform can be used to analyze structures of a periodic nature, such as our dual corridor structure from Chap. 1. Therefore, Fourier methods can be useful for deriving formulae related to corridor paths. In fact, our methods provide a way to encode the *reflection principle* and certain *inclusion-exclusion* arguments currently used in the combinatorial literature. In order to make this text self-contained, we have included all proofs, either explicitly or by way of exercises. Those who wish to go straight to computation may safely skip the technical proofs. The ideas in this chapter will be extended to higher dimensions starting in Chap. 3.

2.1 Reflections and Transitions

In Chap. 1, the vertex numbers $v_{n,a}^{(h)}(k)$ were defined for $h \in \mathbb{N}$, $a \in \mathcal{C}^{(h)}$, and $(n, k) \in \mathbb{N}_0 \times \mathcal{C}^{(h)}$ (see Definition 1.4). In this section, we regard $v_n(k) = v_{n,a}^{(h)}(k)$ as a function of k, called a **state function** for the corridor, and extend the domain to all $k \in \mathbb{Z}$. That is, $v_n : \mathbb{Z} \to \mathbb{Z}$. Henceforth we let $x \in \mathbb{Z}$ denote *position*, and so $v_n(x)$ is the vertex number at position x at *step* (or *time*) $n \in \mathbb{N}_0$.

Recall, the "trick" that helped us compute the infinite corridor numbers (i.e., the number of corridor paths in $\mathbb{N}_0 \times \mathbb{N}$) was to extend the domain of definition of $v_n^{(\infty)}$ to $\mathbb{N}_0 \times \mathbb{Z}$ by *antisymmetry*. A similar technique will be employed for the finite corridor numbers: we require that each v_n must be *admissible*, in a sense that will be defined presently.

From now on, we fix the height $h \in \mathbb{N}$ and let $d = h + 1$. Using d instead of h simplifies our formulas throughout. Consider two **reflections** ρ and σ defined in $\mathbb{Z}$ by:

© Springer Nature Switzerland AG 2019
S. Ault and C. Kicey, *Counting Lattice Paths Using Fourier Methods*, Applied and Numerical Harmonic Analysis,
https://doi.org/10.1007/978-3-030-26696-7_2

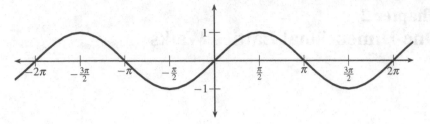

Fig. 2.1 The function $f(x) = \sin(x)$ satisfies $f(-x) = -f(x)$ (antisymmetry), and $f(x + 2\pi) = f(x)$ (2π-periodicity)

$$\rho(x) = -x, \tag{2.1}$$
$$\sigma(x) = 2d - x \tag{2.2}$$

Let $\tau = \sigma\rho$ (here and throughout, $fg = f \circ g$ is composition of functions). Observe that $\tau(x) = \sigma(\rho(x)) = \sigma(-x) = 2d + x$ is simply translation by $2d$.

Definition 2.1 A function $v : \mathbb{Z} \to \mathbb{Z}$ is called **admissible** if v satisfies: $v\rho = -v$, and $v\sigma = -v$.

In other words, v is admissible if and only if the following two conditions are met for all $x \in \mathbb{Z}$.

$$v(-x) = -v(x),$$
$$v(2d - x) = -v(x)$$

Three important observations about admissible functions v are collected below.

1. Because $v\rho = -v$, the function v must be *antisymmetric* (i.e. *odd*) with respect to the origin, and so $v(0) = 0$.
2. Because $v\sigma = -v$ implies that $v(d) = v(2d - d) = -v(d)$, we have $v(d) = 0$.
3. Because $v\tau = v\sigma\rho = -v\rho = v$, v is *2d-periodic*.

For an example of an antisymmetric periodic function that you should be quite familiar with, just consider the sine function, $y = \sin x$, graphed in Fig. 2.1. Of course $y = \sin x$ is defined over all real numbers, in contrast to our functions defined only over the integers.

Now if v_n is a state function for the (h)-corridor (i.e., $\mathcal{C}^{(d-1)}$), we will extend the domain of v_n to all $x \in \mathbb{Z}$ by antisymmetry with respect to the two reflections ρ and σ. That is, extend v_n so that it is admissible.

The action of these reflections is illustrated in Fig. 2.2. Then, with an **initial state** v_0 suitably defined as an admissible function, each state v_{n+1} is obtained from v_n by the following rule:

$$v_{n+1}(x) = v_n(x - 1) + v_n(x + 1) \tag{2.3}$$

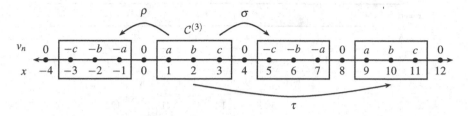

Fig. 2.2 A typical state function describing the dual corridor structure for $d = 4$ ($h = 3$). Note that v_n is defined over all $x \in \mathbb{Z}$. The fundamental region $\mathcal{C}^{(3)}$ consists of the points $\{1, 2, 3\}$

It is straightforward to show that if v_n is admissible, then so is the function v_{n+1} defined by (2.3); the proof of a more general result is given later in Lemma 3.4, but see also Exercise 2.1. In particular, each state v_n has "zero padding" at $x = 0$ and $x = d$. In other words, the recursive formula (2.3) correctly counts corridor vertex numbers even at the upper and lower "boundary" of the corridor:

$$v_{n+1}(1) = v_n(0) + v_n(2) = v_n(2),$$
$$v_{n+1}(d-1) = v_n(d-2) + v_n(d) = v_n(d-2)$$

Hence, simply by redefining the states so that they are admissible, the recursive formula (1.2) given in §1.2 reduces to the much simpler formula given by (2.3). We call the entire array of values $v_n(x)$ for $(n, x) \in \mathbb{N}_0 \times \mathbb{Z}$ (where each v_n is admissible) the **dual corridor structure**, and the subset of lattice points $\mathbb{N}_0 \times \mathcal{C}^{(d-1)} \subset \mathbb{N}_0 \times \mathbb{Z}$ may be referred to as the **fundamental corridor**. (We still say that $\mathcal{C}^{(d-1)}$ is the **fundamental region** for the corridor, as in Chap. 1.) Figure 2.3 illustrates how the state vectors progress recursively within the dual corridor structure.

Now because (2.3) is valid for all $x \in \mathbb{Z}$ and has such a simple form as a sum of shifts of previous states, one can use a *linear operator* to encode the formula. First let us recall a few facts about operators. By **operator**, we mean a function from a vector space[1] to itself. Usually the vector spaces under consideration consist of functions. The set of all functions from a set A to another set B is called a **function space**, and is denoted B^A. If $v : A \to B$ is a function, and if T is an operator on the function space B^A, then $T[v]$ is also a function in B^A. That is, $T[v] : A \to B$, and the output of $T[v]$ on any given $x \in A$ would be written $T[v](x) \in B$. If U is another operator on the same function space, then the composition TU is defined by $(TU)[v] = T[U[v]]$. Our work involves primarily **linear** operators. By definition, a linear operator T satisfies the following properties.

1. $T[u + v] = T[u] + T[v]$, for all functions u, v.
2. $T[cu] = cT[u]$, for all functions u and scalars c (we may assume $c \in \mathbb{C}$).

---

[1] We assume an elementary knowledge of linear algebra, including the definition of a vector space. For review, any undergraduate text in linear algebra would suffice, e.g., [39, 40, 48].

	0	1	2	3	4	5	6	7	8	9	10	11	12
$x=6$													
$x=5$				1		4		13		40		121	
$x=4$			1		4		13		40		121		364
$x=3$		1		3		9		27		81		243	
$x=2$	1		2		5		14		41		122		365
$x=1$		1		2		5		14		41		122	
$x=0$													
$x=-1$		-1		-2		-5		-14		-41		-122	
$x=-2$	-1		-2		-5		-14		-41		-122		-365
$x=-3$		-1		-3		-9		-27		-81		-243	
$x=-4$			-1		-4		-13		-40		-121		-364
$x=-5$				-1		-4		-13		-40		-121	
$x=-6$													
$n=$	0	1	2	3	4	5	6	7	8	9	10	11	12

Fig. 2.3 The dual corridor structure for $d = 6$ ($h = 5$) with initial state corresponding to $a = 2$. The table shows the fundamental corridor together with one negative reflection; this pattern repeats in blocks of 12 up and down. Column n is associated with the sequence $\left(v_{n,2}^{(5)}(x)\right)_{x \in \mathbb{Z}}$, with only nonzero values of v_n indicated in this table. Each corridor number c_n is found by summing the numbers in the nth column in the fundamental region, so we have $(c_{n,2}^{(5)})_{n \in \mathbb{N}_0} = (1, 2, 3, 6, 9, 18, 27, 54, 81, 162, 243, 486, 729, \ldots)$

Although our state functions from Chap. 1 have range in $\mathbb{Z}$, it will become convenient to consider complex-valued functions as well. Because our functions typically take input in $\mathbb{Z}$, we often regard a corridor state function $v : \mathbb{Z} \to \mathbb{C}$ as a sequence $(v(x))_{x \in \mathbb{Z}} = (\ldots, v(-2), v(-1), v(0), v(1), v(2), \ldots)$. We define the following operators that act on functions (sequences) $v \in \mathbb{C}^{\mathbb{Z}}$.

- **Right shift**: $R[v](x) = v(x - 1)$.
- **Left shift**: $L[v](x) = v(x + 1)$.
- **Identity**: $I[v] = v$.

It is advantageous to define a general **shift operator**, R^j for $j \in \mathbb{Z}$:

$$R^j[v](x) = v(x - j)$$

Then we have $R^1 = R$, $R^0 = I$, and $R^{-1} = L$. Now (2.3) can be written more compactly. Let $T = R + L$, which we call a **transition operator**.[2]

---

[2]Specifically, $T = R + L$ will be called the *classical* or *two-way* transition operator.

Table 2.1 Various operators applied to the sequence $v = (\ldots, 0, 0, 3, 0, \mathbf{1}, -4, 7, 0, 0, \ldots)$

	x:	$\cdots$	-4	-3	-2	-1	0	1	2	3	4	$\cdots$
Identity:	$I[v] = v$	$\cdots$	0	0	3	0	**1**	-4	7	0	0	$\cdots$
Right shift:	$R[v]$	$\cdots$	0	0	0	3	**0**	1	-4	7	0	$\cdots$
Left shift:	$R^{-1}[v] = L[v]$	$\cdots$	0	3	0	1	-4	7	0	0	0	$\cdots$
Two-way transition:	$T = R + L$	$\cdots$	0	3	0	4	-4	8	-4	7	0	$\cdots$
Three-way transition:	$T_M = R + I + L$	$\cdots$	0	3	3	4	-3	4	3	7	0	$\cdots$

$$v_{n+1} = R[v_n] + L[v_n] = T[v_n] \tag{2.4}$$

The form of T is directly related to the set of valid moves for two-way corridor paths, up-right and down-right, which may be encoded by the integers 1 and -1, respectively. If we had instead wanted to count three-way paths, the valid moves would be 1 (up-right), 0 (right), and -1 (down-right), and so the corresponding transition operator would take the following form:

$$T_M = R^1 + R^0 + R^{-1} = R + I + L \tag{2.5}$$

Thus, if $v_n(x)$ is the admissible state function that computes the n-length three-way corridor numbers, then the (admissible) state function for those of length $n + 1$ is given by the formula:

$$v_{n+1} = R[v_n] + I[v_n] + L[v_n] = T_M[v_n] \tag{2.6}$$

Compare (2.6) to the recursive formula (1.5).

Remark 2.1 In signal processing jargon, v_{n+1} is given by convolution of v_n with a finite impulse response filter [46].

- The filter $(\ldots, 0, 0, 1, \mathbf{0}, 1, 0, 0, \ldots)$ corresponds to the two-way transition T.
- The filter $(\ldots, 0, 0, 1, \mathbf{1}, 1, 0, 0, \ldots)$ corresponds to the three-way transition T_M.

Example 2.1 Let $v = (\ldots, 0, 0, 3, 0, \mathbf{1}, -4, 7, 0, 0, \ldots) \in \mathbb{C}^{\mathbb{Z}}$. Table 2.1 shows the effect of some of the above operators on v.

2.2 The Discrete Fourier Transform

There is a simple Laurent polynomial generating function for the infinite corridor vertex numbers (see Exercise 1.12). However, it turns out that Laurent polynomials are no longer useful as generating functions for the (h)-corridor vertex numbers for finite h. Instead, when a function is circular or periodic in nature, various Fourier techniques may be useful. Indeed, the complex exponential[3] $e^{i\theta}$ and especially the nth **roots of unity**, that is, the complex solutions to the equation $z^n = 1$,

$$1 = e^0, \ e^{2\pi i/n}, \ e^{4\pi i/n}, \ e^{6\pi i/n}, \ \dots, \ e^{2(n-1)\pi i/n}, \ e^{2n\pi i/n} = 1,$$

may be used to construct generating sequences for n-periodic sequences, a technique that crops up in various places in the literature.[4] The corridor numbers indeed arise from periodic structures; the dual corridor construction is $2d$-periodic for a fixed $d \geq 2$. Moreover, corridor numbers are intimately connected to periodic versions of the Pascal triangle, the so-called **circular Pascal arrays** [6, 35]. In a strong sense, the use of the discrete Fourier series makes this connection explicit and provides a natural alternative to the more traditional enumerative methods for lattice path counting, while at the same time opening the door to extensions and generalizations in higher dimensions and related situations.

Fourier transforms may be defined on a variety of function and sequence spaces, where a function or **signal** on the time side t is analyzed based on sines and cosines (or the complex exponential via Euler's Formula) at various frequencies ω. We focus on what is necessary for our purposes, a variation of the classic finite or discrete Fourier transform. For more background on Fourier transforms from a signal processing point of view, we recommend [46] or [55], much of the latter being freely available online.

Fix $N \in \mathbb{N}$, and consider functions $u : \mathbb{Z} \to \mathbb{C}$ that are N-periodic; that is, $u(x + kN) = u(x)$ for all $k \in \mathbb{Z}$. As above, we typically regard u as a sequence $(u(x))_{x \in \mathbb{Z}} \in \mathbb{C}^{\mathbb{Z}}$. There is an obvious vector space structure on the set of N-periodic functions, allowing us to consider u to be a complex vector. We define the (continuous) *Fourier transform* of u by

$$\mathcal{F}[u](\omega) = U(\omega) = \sum_{x=0}^{N-1} u(x)e^{-\frac{2\pi i}{N}x\omega}, \tag{2.7}$$

where $\omega \in \mathbb{R}$. Note that $U(\omega)$ may be computed using any period of u, that is,

$$U(\omega) = \sum_{x=x_0}^{x_0+N-1} u(x)e^{-\frac{2\pi i}{N}x\omega}, \tag{2.8}$$

for any $x_0 \in \mathbb{Z}$. Furthermore, $U(\omega)$ is also N-periodic, that is, $U(\omega + kN) = U(\omega)$ for all $\omega \in \mathbb{R}$ and $k \in \mathbb{Z}$. The **discrete Fourier transform** or **DFT** of u is simply the restriction of (2.7) to integer values ω.

Next consider an operator $\mathcal{F}^{-1}$ defined on N-periodic complex sequences by the following formula.

---

[3] See Appendix A for a refresher on complex numbers and the complex exponential function.
[4] For a recent explanation and use of this technique, see [49].

$$\mathcal{F}^{-1}[U](x) = u(x) = \frac{1}{N} \sum_{\omega=0}^{N-1} U(\omega) e^{\frac{2\pi i}{N} \omega x} \quad \text{for} \quad x \in \mathbb{Z} \qquad (2.9)$$

Here, we interpret the domain of the function u as restricted to integer values: $u = (u(x))_{x \in \mathbb{Z}}$, or $u = (u(x))_{x=x_0}^{x_0+N-1}$. In Exercise 2.6 you will be asked to prove that $\mathcal{F}$ is invertible with inverse $\mathcal{F}^{-1}$ given by (2.9). Note that both $\mathcal{F}$ and $\mathcal{F}^{-1}$ are linear operators.

Example 2.2 Consider the period four sequence $u = (\ldots, \mathbf{2}, 3, 1, -4, \ldots)$. We will find the values of $\mathcal{F}[u](\omega) = U(\omega)$ for $\omega \in \{0, 1, 2, 3\}$. Note that $e^{-\frac{2\pi i}{4}} = -i$, so (2.7) yields:

$$U(\omega) = \sum_{x=0}^{3} u(x) e^{-\frac{2\pi i}{4} x \omega} = \sum_{x=0}^{3} u(x)(-i)^{x\omega}$$

Expanding further, we may write:

$$U(\omega) = u(0) + u(1)(-i)^{\omega} + u(2)(-1)^{\omega} + u(3)i^{\omega}$$
$$= 2 + 3(-i)^{\omega} + (-1)^{\omega} - 4i^{\omega}$$

Hence,

$$U(0) = 2 + 3 + 1 - 4 = 2$$
$$U(1) = 2 - 3i - 1 - 4i = 1 - 7i$$
$$U(2) = 2 - 3 + 1 + 4 = 4$$
$$U(3) = 2 + 3i - 1 + 4i = 1 + 7i$$

Now let's check that the inverse DFT of $U = (2, 1 - 7i, 4, 1 + 7i)$ is in fact equal to u.

$$\mathcal{F}^{-1}[U](x) = \frac{1}{4} \sum_{\omega=0}^{3} U(\omega) e^{\frac{2\pi i}{4} \omega x}$$

$$= \frac{1}{4} \left[U(0) + U(1)i^x + U(2)(-1)^x + U(3)(-i)^x \right]$$

$$= \frac{1}{4} \left[2 + (1 - 7i)i^x + 4(-1)^x + (1 + 7i)(-i)^x \right]$$

It is easily verified that $(\mathcal{F}^{-1}[U])_{x=0}^{3} = (2, 3, 1, -4)$ as expected.

Recall from Sect. 2.1 that the right shift operator R satisfies $R[u](x) = v(x - 1)$ for all $x \in \mathbb{Z}$, where $u \in \mathbb{C}^{\mathbb{Z}}$ is a function. Now suppose that u is N-periodic. The relation between the corresponding Fourier transforms, $\mathcal{F}[R[u]]$ and $\mathcal{F}[u]$, can be derived very easily. All that it takes is a shift in the index of the sum and factoring out the "extra" exponential that arises.

$$\mathcal{F}\left[R[u]\right] = \sum_{x=0}^{N-1} u(x-1)e^{-\frac{2\pi i}{N}x\omega}$$

$$= \sum_{x=-1}^{N-2} u(x)e^{-\frac{2\pi i}{N}(x+1)\omega}$$

$$= e^{-\frac{2\pi i \omega}{N}} \sum_{x=-1}^{N-2} u(x)e^{-\frac{2\pi i}{N}x\omega}$$

$$= e^{-\frac{2\pi i \omega}{N}} \mathcal{F}[u]$$

It follows that for any $k \in \mathbb{Z}$,

$$\mathcal{F}\left[R^k[u]\right](\omega) = e^{-\frac{2k\pi i \omega}{N}} U(\omega), \tag{2.10}$$

where $U = \mathcal{F}[u]$, and in particular, $\mathcal{F}\left[L[u]\right](\omega) = e^{\frac{2k\pi i \omega}{N}} U(\omega)$. In other words, to shift an N-periodic sequence $u(x)$ to the right or left, all that is required on the Fourier side is to multiply $U(\omega)$ against the Nth roots of unity. This is a kind of "convolution" theorem necessary for our purposes, further illustrated by the following example.

Example 2.3 Suppose the Fourier transform of $u(x)$ of period four is given by the following.

$$(U(\omega))_{\omega=0}^3 = (1, 2, 3, 1)$$

If $T = L + R$ and $u_1 = T[u]$, then by Euler's Identity for cosine (A.5), we have:

$$U_1(\omega) = e^{\frac{2\pi i \omega}{4}} U(\omega) + e^{-\frac{2\pi i \omega}{4}} U(\omega) = 2\cos\left(\frac{\pi\omega}{2}\right) U(\omega)$$

One period of U_1 is displayed below.

$$(U_1(\omega))_{\omega=0}^3 = (2 \cdot U(0), 0 \cdot U(1), -2 \cdot U(2), 0 \cdot U(3)) = (2, 0, -6, 0)$$

In the previous example, if $u_n = T^n[u]$, then a straightforward induction shows that $U_n(\omega) = \left(2\cos(\frac{\pi\omega}{2})\right)^n U(\omega)$. Thus repeated composition by the operator T becomes repeated multiplication on the Fourier side. By abuse of standard Fourier notation, we may write $U_n = \widehat{T}^n U$, where $\widehat{T}(\omega) = 2\cos(\frac{\pi\omega}{2})$; see also the construction of $\widehat{T}(\omega)$ in Lemma 3.6. In fact, this relationship between composition and multiplication is exactly what allows us to construct explicit formulae for vertex states, as we shall explain in what follows.

Next, let us define a very simple function that will aid us in developing the rest of the theory. Define the **delta function** $\delta : \mathbb{Z} \to \mathbb{Z}$ as follows:

$$\delta(x) = \begin{cases} 1 & x = 0, \\ 0 & x \neq 0 \end{cases} \tag{2.11}$$

Clearly δ is not periodic. However we may proceed to find its DFT as if it were periodic (this is a standard trick in Fourier analysis). More precisely, we take the DFT of a *periodization* of δ. Let $N \geq 1$ and consider the related function:

$$\tilde{\delta}(x) = \begin{cases} 1 & x = kN \text{ for some } k \in \mathbb{Z}, \\ 0 & \text{otherwise} \end{cases} \tag{2.12}$$

Certainly $\tilde{\delta}(x) = \delta(x)$ for all $x \in \{0, 1, \ldots, N-1\}$, and because the DFT is defined as a sum over x in a single periodic block, we may *define* the DFT of δ to be $\mathcal{F}[\tilde{\delta}]$. This is merely a technical detail in order to stay consistent with definitions; henceforth, we have no need for explicit mention of $\tilde{\delta}$. The DFT of δ is especially simple:

$$\mathcal{F}[\delta](\omega) = \sum_{x=0}^{N-1} \delta(x) e^{-\frac{2\pi i}{N} x \omega} = (1)e^0 + (0)e^{-\frac{2\pi i}{N} \omega} + \cdots + (0)e^{-\frac{2(N-1)\pi i}{N} \omega} = 1 \tag{2.13}$$

Now let $a \in \mathbb{Z}$, and define $\delta_a(x) = \delta(x - a)$. Equivalently, $\delta_a = R^a[\delta]$. For our purposes, we require $0 \leq a \leq N$, and define $\mathcal{F}[\delta_a]$ as the DFT of the N-periodization of δ_a. Using (2.10) with (2.13), we have the following formula:

$$\mathcal{F}[\delta_a] = \mathcal{F}\left[R^a[\delta]\right] = e^{-\frac{2a\pi i \omega}{N}} \tag{2.14}$$

We require an admissible version of δ_a, which we denote by Δ_a. In particular, Δ_a must be $2d$-periodic and antisymmetric with respect to the origin. Fix $1 \leq a \leq d - 1$ and define $\Delta_a(x)$ for $0 \leq x < N$ as follows.

$$\Delta_a(x) = \delta_a(x) - \delta_{-a}(x) \tag{2.15}$$

Then extend the domain of Δ_a to all of $\mathbb{Z}$ by $2d$-periodicity. The DFT of Δ_a (which depends on the chosen value of d) is derived below. The last equality requires Euler's Identity for sine (A.5).

$$\mathcal{F}[\Delta_a] = \mathcal{F}[\delta_a(x)] - \mathcal{F}[\delta_{-a}(x)] = e^{-\frac{2a\pi i \omega}{N}} - e^{\frac{2a\pi i \omega}{N}} = -2i \sin \frac{2a\pi \omega}{N} \tag{2.16}$$

Example 2.4 Suppose $d = 4$. Below is the $2d$-periodic version of Δ_3.

$$(\Delta_3(x))_{x=-4}^3 = (\ldots, 0, -1, 0, 0, \mathbf{0}, 0, 0, 1, \ldots)$$

The continuous Fourier transform of the above is: $-2i \sin(\frac{3\pi \omega}{4})$.

2.3 Computing Vertex Numbers Using the DFT

We now use Fourier methods to derive some trigonometric formulae related to corridors and lattice paths. For now let's assume classical (two-way) corridor paths. Let $d \geq 2$, $a \in \mathcal{C}^{(d-1)}$, and suppose that $v_n = v_{n,a}^{(d-1)}$ is an admissible state function that counts corridor paths of length n in $\mathbb{N}_0 \times \mathcal{C}^{(d-1)}$. In particular, $v_n(x)$ is $2d$-periodic and antisymmetric with respect to the origin. As discussed above, $v_{n+1} = T[v_n]$, where $T = R + L$, would then be the admissible state function that counts corridor paths of length $n + 1$. Let's establish the relationship between the Fourier transforms of the two states V_n and V_{n+1}. First, by definitions and linearity of the Fourier transform, we find:

$$V_{n+1} = \mathcal{F}[v_{n+1}] = \mathcal{F}[T[v_n]] = \mathcal{F}[R[v_n] + L[v_n]] = \mathcal{F}[R[v_n]] + \mathcal{F}[L[v_n]]$$

Next, making use of (2.10) with $N = 2d$ and Euler's Identity for cosine (A.5) we obtain the following formula.

$$V_{n+1}(\omega) = e^{-\frac{\pi i \omega}{d}} \mathcal{F}[v_n](\omega) + e^{\frac{\pi i \omega}{d}} \mathcal{F}[v_n](\omega) = 2\cos\left(\frac{\pi\omega}{d}\right) V_n(\omega) \qquad (2.17)$$

Hence by induction,

$$V_n(\omega) = \left(2\cos\left(\frac{\pi\omega}{d}\right)\right)^n V_0(\omega), \qquad (2.18)$$

where $V_0 = \mathcal{F}[v_0]$, and v_0 is the admissible initial state function corresponding to the initial point $a \in \mathcal{C}^{(d-1)}$. In particular, $v_0(a) = 1$ and $v_0(x) = 0$ for $x \neq a$ within the fundamental region. This implies that $v_0(x)$ coincides with $\delta_a(x)$ for $1 \leq x \leq d - 1$. Recall that $\Delta_a(x)$ as defined by (2.15) also coincides with $\delta_a(x)$ for $1 \leq x \leq d - 1$ and has the added benefit of being antisymmetric and $2d$-periodic—i.e., Δ_a is admissible. Therefore, with the help of (2.16), we obtain:

$$V_0(\omega) = \mathcal{F}[\Delta_a] = -2i\sin\left(\frac{\pi\omega a}{d}\right) \qquad (2.19)$$

Taking this result together with (2.18), the formula for V_n follows.

$$V_n(\omega) = \left(2\cos\left(\frac{\pi\omega}{d}\right)\right)^n V_0(\omega) = -2^{n+1}i\cos^n\left(\frac{\pi\omega}{d}\right)\sin\left(\frac{\pi\omega a}{d}\right) \qquad (2.20)$$

Now we have all of the ingredients to prove the main theorem.

Theorem 2.1 *Let $h \in \mathbb{N}$ and $d = h + 1$. The vertex numbers for n-length corridor paths in the (h)-corridor, with initial point $(0, a)$, are given by:*

$$v_{n,a}(x) = \frac{2^n}{d} \sum_{\omega=0}^{2d-1} \sin\left(\frac{\pi\omega x}{d}\right)\cos^n\left(\frac{\pi\omega}{d}\right)\sin\left(\frac{\pi\omega a}{d}\right) \qquad (2.21)$$

proof Apply the inverse Fourier transform to $V_n(\omega)$ as given by (2.20).

$$v_{n,a}(x) = \frac{1}{2d} \sum_{\omega=0}^{2d-1} V_n(\omega) e^{\frac{\pi i}{d}\omega x} \tag{2.22}$$

Noting that $V_n(\omega)$ is purely imaginary and using Euler's Formula (A.4), we isolate the real part of the summand because $v_{n,a}(x)$ must be real (indeed, in $\mathbb{Z}$).

$$
\begin{aligned}
V_n(\omega) e^{\frac{\pi i}{d}\omega x} &= V_n(\omega) \left[\cos\left(\frac{\pi \omega x}{d}\right) + i \sin\left(\frac{\pi \omega x}{d}\right) \right] \\
&= i \sin\left(\frac{\pi \omega x}{d}\right) V_n(\omega) \\
&= i \sin\left(\frac{\pi \omega x}{d}\right) (-2^{n+1}i) \cos^n\left(\frac{\pi \omega}{d}\right) \sin\left(\frac{\pi \omega a}{d}\right) \\
&= 2^{n+1} \sin\left(\frac{\pi \omega x}{d}\right) \cos^n\left(\frac{\pi \omega}{d}\right) \sin\left(\frac{\pi \omega a}{d}\right)
\end{aligned}
$$

Finally, putting this expression back into (2.22) completes the proof. □

By inherent symmetry, less than *half* of the sum in (2.21) is needed. Indeed, the terms corresponding to ω and $2d - \omega$ are exactly the same, and those corresponding to $\omega = 0$ and $\omega = d$ are zero, so we may write:

$$v_{n,a}(x) = \frac{2^{n+1}}{d} \sum_{\omega=1}^{d-1} \sin\left(\frac{\pi \omega x}{d}\right) \cos^n\left(\frac{\pi \omega}{d}\right) \sin\left(\frac{\pi \omega a}{d}\right) \tag{2.23}$$

Remark 2.2 Formula (2.23) is equivalent to Eq. (5) of Krattenthaler and Mohanty [37], after the simple change of variables, $d \mapsto K+2$, $\omega \mapsto k$, $n \mapsto \ell$, $a \mapsto r+1$, and $x \mapsto s+1$. See also Theorem 1 of Felsner and Heldt [25].

The corresponding corridor numbers are given by:

Corollary 2.1 *Let $h \in \mathbb{N}$ and $d = h+1$. The corridor numbers for n-length corridor paths in the (h)-corridor, with initial point $(0, a)$, are given by:*

$$c_{n,a} = \frac{2^{n+1}}{d} \sum_{\xi=0}^{\lfloor \frac{d}{2}\rfloor-1} \left[1 + \cos\left(\frac{\pi(2\xi+1)}{d}\right)\right] \cos^n\left(\frac{\pi(2\xi+1)}{d}\right) \frac{\sin\left(\frac{\pi(2\xi+1)a}{d}\right)}{\sin\left(\frac{\pi(2\xi+1)}{d}\right)} \tag{2.24}$$

proof By definition, $c_{n,a} = \sum_{x=0}^{d-1} v_n(x)$, and by (2.22), we have:

$$c_{n,a} = \frac{1}{2d} \sum_{\omega=0}^{2d-1} V_n(\omega) \sum_{x=0}^{d-1} e^{i\pi\omega x/d} \tag{2.25}$$

Note if $\omega = 0$, then $V_n(\omega)$ vanishes, else we sum the geometric series and simplify.

$$\sum_{x=0}^{d-1} \left(e^{i\pi\omega/d}\right)^x = \frac{1 - e^{i\pi\omega}}{1 - e^{i\pi\omega/d}}$$

$$= \frac{1 - (-1)^\omega}{1 - \cos\left(\frac{\pi\omega}{d}\right) - i\sin\left(\frac{\pi\omega}{d}\right)}$$

$$= \frac{1 - (-1)^\omega}{2\left[1 - \cos\left(\frac{\pi\omega}{d}\right)\right]} \left[1 - \cos\left(\frac{\pi\omega}{d}\right) + i\sin\left(\frac{\pi\omega}{d}\right)\right]$$

Note that only the imaginary part of this expression will be needed since it is to be inserted into (2.25), as $V_n(\omega)$ itself is purely imaginary (as always, we know a priori that the corridor numbers must be real) and use the familiar trigonometric identity, $\frac{\sin(\theta)}{1-\cos(\theta)} = \frac{1+\cos(\theta)}{\sin(\theta)}$.

$$\text{Im}\left(\sum_{x=0}^{d-1} \left(e^{i\pi\omega/d}\right)^x\right) = \frac{1 - (-1)^\omega}{2} \cdot \frac{\sin\left(\frac{\pi\omega}{d}\right)}{1 - \cos\left(\frac{\pi\omega}{d}\right)} = \frac{1 - (-1)^\omega}{2} \cdot \frac{1 + \cos\left(\frac{\pi\omega}{d}\right)}{\sin\left(\frac{\pi\omega}{d}\right)}$$

Finally, making the change of variables $\omega = 2\xi + 1$, the result follows. □

When $a = 1$, (2.24) simplifies.

$$c_n = \frac{2^{n+1}}{d} \sum_{\xi=0}^{\lfloor\frac{d}{2}\rfloor-1} \left[1 + \cos\left(\frac{\pi(2\xi+1)}{d}\right)\right]\cos^n\left(\frac{\pi(2\xi+1)}{d}\right) \qquad (2.26)$$

Example 2.5 The corridor numbers for $d = 5$ ($h = 4$) and $a = 1$ provide an interesting link to the well-known formula for the Fibonacci numbers.

$$F_n = c_n^{(4)}$$

$$= \frac{2^{n+1}}{5} \sum_{\xi=0}^{2} \left[1 + \cos\left(\frac{\pi(2\xi+1)}{5}\right)\right]\cos^n\left(\frac{\pi(2\xi+1)}{5}\right)$$

$$= \frac{2^{n+1}}{5} \left(\left[1 + \cos\left(\frac{\pi}{5}\right)\right]\cos^n\left(\frac{\pi}{5}\right) + \left[1 + \cos\left(\frac{3\pi}{5}\right)\right]\cos^n\left(\frac{3\pi}{5}\right) + (1 + (-1))(-1)^n\right)$$

$$= \frac{5 + \sqrt{5}}{10}\left(\frac{1 + \sqrt{5}}{2}\right)^n + \frac{5 - \sqrt{5}}{10}\left(\frac{1 - \sqrt{5}}{2}\right)^n$$

2.4 Application: Three-Way Paths

With little additional effort we extend the methods of Sect. 2.3 to count the number of three-way corridor paths. We will continue to use the notation v_n for the state vector in order to highlight how similar the methods are in comparison to the classical states (whereas in Chap. 1, the notation for three-way state vectors took the form m_n). The three-way transition operator is $T_M = L + I + R$.

$$v_{n+1} = T_M[v_n] = R[v_n] + v_n + R^{-1}[v_n]$$
$$V_{n+1} = \mathcal{F}[v_{n+1}] = \mathcal{F}[R[v_n]] + \mathcal{F}[v_n] + \mathcal{F}[R^{-1}[v_n]]$$

Therefore, we derive the following relationship between V_{n+1} and V_n.

$$V_{n+1}(\omega) = e^{-\frac{i\pi\omega}{d}} V_n(\omega) + V_n(\omega) + e^{\frac{i\pi\omega}{d}} V_n(\omega) = \left[1 + 2\cos\left(\frac{\pi\omega}{d}\right)\right] V_n(\omega) \quad (2.27)$$

Thus with initial point $a \in C^{(d-1)}$ for our corridor paths, we may use (2.19) to obtain the following.

$$V_n(\omega) = \left[1 + 2\cos\left(\frac{\pi\omega}{d}\right)\right]^n V_0(\omega) = -2i\left[1 + 2\cos\left(\frac{\pi\omega}{d}\right)\right]^n \sin\left(\frac{\pi\omega a}{d}\right)$$
$$(2.28)$$

One can now obtain the analog of Theorem 2.1 by taking an inverse DFT. Specifically the number of three-way corridor paths beginning at $(0, a)$ in $\mathbb{N}_0 \times C^{(d-1)}$ is given by:

$$v_{n,a}(x) = \frac{1}{d} \sum_{\omega=0}^{2d-1} \sin\left(\frac{\pi\omega x}{d}\right) \left[1 + 2\cos\left(\frac{\pi\omega}{d}\right)\right]^n \sin\left(\frac{\pi\omega a}{d}\right) \quad (2.29)$$

Then observe that (2.29) may be condensed, as we have done in (2.23), to obtain:

$$v_{n,a}(x) = \frac{2}{d} \sum_{\omega=1}^{d-1} \sin\left(\frac{\pi\omega x}{d}\right) \left[1 + 2\cos\left(\frac{\pi\omega}{d}\right)\right]^n \sin\left(\frac{\pi\omega a}{d}\right) \quad (2.30)$$

Remark 2.3 The same change of variables as before, $d \mapsto K + 2$, $\omega \mapsto k$, $n \mapsto \ell$, $a \mapsto r + 1$, and $x \mapsto s + 1$, converts (2.30) into Formula (6) of [37].

Furthermore, if $M_{n,a}$ stands for the total number of three-way paths of length n, beginning at $(0, a)$, then we obtain the analog of Corollary 2.1:

$$M_{n,a} =$$

$$\frac{2}{d} \sum_{\xi=0}^{\lfloor \frac{d}{2} \rfloor - 1} \left[1 + \cos\left(\frac{\pi(2\xi + 1)}{d}\right)\right] \left[1 + 2\cos\left(\frac{\pi(2\xi + 1)}{d}\right)\right]^n \frac{\sin\left(\frac{\pi(2\xi+1)a}{d}\right)}{\sin\left(\frac{\pi(2\xi+1)}{d}\right)}$$
$$(2.31)$$

Remark 2.4 There is an interesting link between two-way and three-way corridor paths that does not directly involve Fourier methods but instead has to do with how their transition operators behave. For clarity here we will revert back to the notation $m_{n,a}(x)$ for the three-way vertex numbers, while $v_{n,a}(x)$ stands for the two-way vertex numbers. Then there is a relation, which you are asked to prove in Exercise 2.12:

$$m_{n,a}(x) = \sum_{k=0}^{n} \binom{n}{k} v_{n,a}(x) \tag{2.32}$$

2.5 Application: Unbounded Corridors

We will now use our Fourier methods to develop an integral formula for the number of two-way lattice paths in the infinite corridor $\mathbb{N}_0 \times \mathcal{C}^{(\infty)} = \mathbb{N}_0 \times \mathbb{N}$ (recall Sect. 1.4). Fix an initial point $a \in \mathcal{C}^{(\infty)}$. Observe, if $h \geq n + a$, then $v_{n,a}^{(\infty)}(x) = v_{n,a}^{(h)}(x)$ for every $x \in \mathbb{N}$, since no path of length n can cross the upper boundary of $\mathcal{C}^{(h)}$. Thus $v_n^{(h)}$ stabilizes for high enough h, and so we define the infinite vertex and corridor numbers as follows.

$$v_n^{(\infty)}(x) = \lim_{h \to \infty} v_n^{(h)}(x), \quad \text{and} \quad c_n^{(\infty)}(x) = \lim_{h \to \infty} c_n^{(h)}(x)$$

As before, we let $d = h + 1$ in order to simplify formulae going forward. Now consider n and x as parameters (constants) and define a function of $t \in \mathbb{R}$ as follows:

$$f_{n,x,a}(t) = \sin(tx) \cos^n(t) \sin(ta) \tag{2.33}$$

Observe that $f_{n,x,a}(\pi) = 0$. Equation (2.33) satisfies the following equation which links it directly to our vertex numbers via Formula (2.23).

$$\frac{2^{n+1}}{d} \sum_{k=1}^{d} f_{n,x,a}\left(\frac{k\pi}{d}\right) = \frac{2^{n+1}}{d} \sum_{k=1}^{d} \sin\left(\frac{k\pi x}{d}\right) \cos^n\left(\frac{k\pi}{d}\right) \sin\left(\frac{k\pi a}{d}\right) = v_{n,a}^{(d-1)}(x) \tag{2.34}$$

In fact Eq. (2.34) has the form of a *Riemann sum*. We will integrate (2.33) on the interval $[0, \pi]$. From basic calculus, we know that the value of a definite integral is equal to the limit of the Riemann sum (assuming the integrand is integrable) [60, 63]. Consider d equal subintervals of $[0, \pi]$ and a right endpoints Riemann sum as suggested by Fig. 2.4.

With width $\Delta t = \frac{\pi}{d}$, and right endpoints $t_k = \frac{k\pi}{d}$, the Riemann sum can easily be set up, and taking the limit as $d \to \infty$, we find:

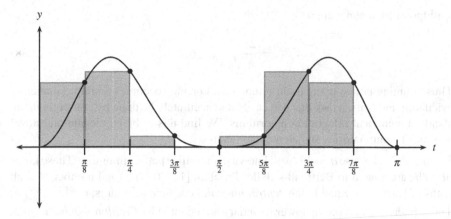

Fig. 2.4 A plot of $f_{3,2,1}(t) = \sin(2t)\cos^3(t)\sin(t)$ and $d = 8$ subintervals for a right Riemann sum

$$\int_0^\pi f_{n,x,a}(t)\,dt = \lim_{d\to\infty} \sum_{k=1}^d f_{n,x,a}(t_k)\,\Delta t$$

$$= \lim_{d\to\infty} \frac{\pi}{d} \sum_{k=1}^d f_{n,x,a}\left(\frac{k\pi}{d}\right)$$

$$= \frac{\pi}{2^{n+1}} \lim_{d\to\infty} v_n^{(d-1)}(x)$$

Therefore, with Eq. (2.34) in mind, we find:

$$v_n^{(\infty)}(x) = \frac{2^{n+1}}{\pi} \int_0^\pi \sin(tx)\cos^n(t)\sin(ta)\,dt \tag{2.35}$$

A similar analysis can be used for the corridor numbers $c_{n,a}^{(\infty)}$. Let us define a new function $g_{n,a}$ based on Formula (2.24).

$$g_{n,a}(x) = (1 + \cos(t))\cos^n(t)\frac{\sin(ta)}{\sin(t)} \tag{2.36}$$

Assume d is even for simplicity. Then by Corollary 2.1 we have:

$$c_{n,a}^{(d-1)} = \frac{2^{n+1}}{d} \sum_{\xi=0}^{\frac{d}{2}-1} g_{n,a}\left(\frac{\pi(2\xi+1)}{d}\right) \tag{2.37}$$

This time, the points $t = \frac{\pi(2\xi+1)}{d}$ are midpoints of subintervals of length $\Delta t = \frac{2\pi}{d}$. There are $\frac{d}{2}$ such subintervals in $[0, \pi]$; hence we obtain our integral formula through a midpoint Riemann sum.

$$c_{n,a}^{(\infty)} = \frac{2^n}{\pi} \int_0^{\pi} g_{n,a}(t)\, dt \tag{2.38}$$

Thus counting lattice paths in an unbounded corridor becomes a simple matter of evaluating integrals, a task that can easily and accurately be done by computer using standard numerical integration algorithms. We find this to be an elegant and novel merger of combinatorics and elementary analysis.

Remark 2.5 The case $a = 1$ is well known in the lattice path community. These kinds of paths are related to Bertrand's Ballot Problem [1, 10]. The total number of such paths of length n is equal to the *central binomial coefficient*,[5] that is, $c_n^{(\infty)} = \binom{n}{\lfloor n/2 \rfloor}$. Those paths that end on the lower boundary are counted by *Catalan numbers*,[6] that is, $v_{2m}^{(\infty)}(1) = C_m$. These results can be recovered from (2.35) and (2.38).

2.6   Application: Walks in Cycle Graphs

Consider the following graph-theoretic problem: *How many walks are there of length n, beginning at a fixed vertex, and ending at some specified vertex in the cycle graph of order m?* We will be able to easily answer this question by reinterpreting the question in terms of paths in a **cylindrical corridor** (see Fig. 2.5).

The key is to construct an appropriate periodic state function $v_n(x)$. Let us consider the classical transition operator $T = R + L$. For convenience in formulae, we may assume that the fundamental region is $\{0, 1, 2, \ldots, m - 1\}$ for some $m \geq 1$, with the understanding that $x = m$ is to be identified with $x = 0$. We may also assume that all paths begin at $x = 0$, as other starting points give the same counts by cyclic symmetry. Now define the initial state as follows:

$$v_0(x) = \begin{cases} 1 & x = km \text{ for some } k \in \mathbb{Z}, \\ 0 & \text{otherwise} \end{cases}$$

Clearly v_0 is periodic[7] of period m and effectively encodes the identification of boundary points $0 \equiv m$. It should be apparent that the recursive rule $v_{n+1}(x) = v_n(x - 1) + v_n(x + 1) = T[v_n](x)$ is valid for every $(n, x) \in \mathbb{N} \times \mathbb{Z}$, hence the state function $v_n(x) = T^n[v_0](x)$ does indeed count the number of paths of length n ending at x in our cylindrical corridor. Unlike in the case of standard corridors, here we do not require *admissibility* (which includes antisymmetry). Instead we call this progression of periodic states the **cyclic corridor** structure (see Fig. 2.6).

---

[5] OEIS sequence A000984.
[6] OEIS sequence A000108 in the OEIS. See also Exercise 1.7.
[7] Indeed, v_0 is the m-periodization of δ as defined in Eq. (2.11).

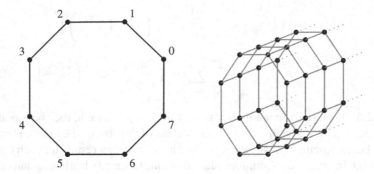

Fig. 2.5 *(Left)* The cycle graph of order $m = 8$ with vertices labelled $0, 1, \ldots 7$. *(Right)* The corresponding cylindrical corridor

Fig. 2.6 The cyclic corridor structure for $m = 6$ as generated by a spreadsheet program. According to the data there are $v_8(0) = 86$ walks of length 8 starting and ending at the same point in the cycle graph of order 6

k	n=0	1	2	3	4	5	6	7	8
17		1	0	3	0	11	0	43	0
16		0	1	0	5	0	21	0	85
15		0	0	2	0	10	0	42	0
14		0	1	0	5	0	21	0	85
13		1	0	3	0	11	0	43	0
12	1	0	2	0	6	0	22	0	86
11		1	0	3	0	11	0	43	0
10		0	1	0	5	0	21	0	85
9		0	0	2	0	10	0	42	0
8		0	1	0	5	0	21	0	85
7		1	0	3	0	11	0	43	0
6	1	0	2	0	6	0	22	0	86
5		1	0	3	0	11	0	43	0
4		0	1	0	5	0	21	0	85
3		0	0	2	0	10	0	42	0
2		0	1	0	5	0	21	0	85
1		1	0	3	0	11	0	43	0
0	1	0	2	0	6	0	22	0	86
-1		1	0	3	0	11	0	43	0
-2		0	1	0	5	0	21	0	85
-3		0	0	2	0	10	0	42	0
-4		0	1	0	5	0	21	0	85
-5		1	0	3	0	11	0	43	0
-6	1	0	2	0	6	0	22	0	86

Because v_0 is m-periodic and the transition operator T preserves this periodicity, the DFT can be used in much the same way as in Theorem 2.1. First we find the DFT of the initial state (but keep in mind that the periodicity is now m rather than $2d$).

$$V_0(\omega) = \sum_{x=0}^{m-1} v_0(x)e^{\frac{-2\pi i}{m}x\omega} = (1)e^0 + (0)e^{\frac{-2\pi i}{m}\omega} + \cdots + (0)e^{\frac{-2\pi i}{m}(m-1)\omega} = 1$$

Then the required state function can be constructed.

$$v_n(x) = \mathcal{F}^{-1}\left[\widehat{T}^n V_0\right] = \frac{1}{m}\sum_{\omega=0}^{m-1} e^{\frac{2\pi i}{m}\omega x}\left(2\cos\left(\frac{2\pi\omega}{m}\right)\right)^n$$

$$= \frac{2^n}{m}\sum_{\omega=0}^{m-1}\cos\left(\frac{2\pi\omega x}{m}\right)\cos^n\left(\frac{2\pi\omega}{m}\right) \qquad (2.39)$$

Remark 2.6 Because there are no boundaries in the cycle graph, the corresponding corridor numbers are trivially $c_n = 2^n$, regardless of the size of the graph. However by using Fourier methods as shown above, we have *stratified* results for each distinct ending point. In other words, our formulas allow one to specify beginning and ending points independently.

However, certain scenarios allow one to simplify Eq. (2.39). For example, when the starting and ending points are the same ($x = 0$), then we obtain:

$$v_n(0) = \frac{2^n}{m}\sum_{\omega=0}^{m-1}\cos^n\left(\frac{2\pi\omega}{m}\right)$$

For each $m \geq 1$, the sequence $(v_n(0))_{n\geq 0}$ records the number of closed walks in the cycle graph of order m, by length.[8]

Example 2.6 The smallest nontrivial case, $m = 2$, represents a graph on two vertices with two edges connecting them. In this case, there are no odd-length paths starting and ending at the same vertex. However, for even n, the number of such paths is a power of 2.

$$v_n(0) = \frac{2^n}{2}\left(1 + \cos^n(\pi\omega)\right) = 2^{n-1}\left(1 + (-1)^n\right) = \begin{cases} 2^n & \text{if } n \text{ is even,} \\ 0 & \text{if } n \text{ is odd} \end{cases}$$

Exercises

2.1 Suppose v is an admissible state, and let $T = R + L$. Prove that $T[v]$ is also admissible.

2.2 Prove the following properties.

(a) $\rho^2(x) = x$
(b) $\sigma^2(x) = x$
(c) $RL = LR = I$
(d) $R^{j+k} = R^j R^k$

2.3 Suppose a function v satisfies antisymmetry and N-periodicity for some $N \in \mathbb{N}$. Show that $v(N - x) = -v(x)$. (Hence, admissibility is equivalent to antisymmetry and $2d$-periodicity.)

---

[8] See OEIS sequences A078008, A199573, A054877, A047849, A094659.

2.4 Find the discrete Fourier transform of each of the following N-periodic sequences (entry corresponding to $x = 0$ bolded in each).

(a) $(\ldots, \mathbf{0}, 1, 0, -1, \ldots)$, $N = 4$
(b) $(\ldots, 0, 1, -3, \mathbf{0}, 3, -1, 0, \ldots)$, $N = 6$
(c) $(\ldots, 0, 1, \mathbf{3}, 1, 0, 1, 2, 1, \ldots)$, $N = 8$
(d) $(\ldots, \mathbf{5}, 0, 2, 3, 0, 0, \ldots)$, $N = 6$

2.5 Compute the inverse DFT of each of your answers to Exercise 2.4, verifying that $\mathcal{F}^{-1}[\mathcal{F}[u]] = u$ in each case.

2.6 Fix $N \in \mathbb{N}$. Prove:

(a) $\mathcal{F}^{-1}[\mathcal{F}[u]] = u$, for all N-periodic functions u.
(b) $\mathcal{F}[\mathcal{F}^{-1}[U]] = U$, for all N-periodic functions U.

2.7 Let x_0 and ω_0 be arbitrary integers. Suppose $u \in \mathbb{C}^{\mathbb{Z}}$ is N-periodic and let $U = \mathcal{F}[u]$. It can be shown that sequences u and U satisfy a version of *Parseval's Theorem*.

$$\sum_{x=x_0}^{x_0+N-1} |u(x)|^2 = \frac{1}{N} \sum_{\omega=\omega_0}^{\omega_0+N-1} |U(\omega)|^2 \tag{2.40}$$

Verify that u and U from Example 2.2 do indeed satisfy (2.40). (We will encounter applications of Parseval's Theorem in Sect. 4.1).

2.8 Suppose that $u \in \mathbb{C}^{\mathbb{Z}}$ is $2d$-periodic and symmetric about the origin (i.e., $u(-x) = u(x)$ for all $x \in \mathbb{Z}$, or u is an **even** function). Prove the following formula and compare your result to Exercise 2.4(c).

$$\mathcal{F}[u](\omega) = u(0) + u(d)(-1)^\omega + 2 \sum_{x=1}^{d-1} u(x) \cos\left(\frac{\pi x \omega}{d}\right) \tag{2.41}$$

2.9 Suppose that $u \in \mathbb{C}^{\mathbb{Z}}$ is $2d$-periodic and antisymmetric about the origin (i.e. $u(-x) = -u(x)$ for all $x \in \mathbb{Z}$, or u is an **odd** function).

(a) Show that $u(qd) = 0$ for all $q \in \mathbb{Z}$.
(b) Prove the following DFT formula for odd $2d$-periodic functions u, comparing your result to Exercises 2.4(a, b).

$$\mathcal{F}[u](\omega) = -2i \sum_{x=1}^{d-1} u(x) \sin\left(\frac{\pi x \omega}{d}\right) \tag{2.42}$$

2.10 Suppose that $u \in \mathbb{C}^{\mathbb{Z}}$ is N-periodic.

(a) Define $a(x) = \frac{1}{2}[u(x) + u(-x)]$, and $b(x) = \frac{1}{2}[u(x) - u(-x)]$. Show that a is even, b is odd, and $u = a + b$.

(b) Let $x_0 \in \mathbb{Z}$ be arbitrary. Show that $\sum_{x=x_0}^{x_0+N-1} a(x)\overline{b(x)} = 0$ (i.e. a and b are *orthogonal* in the function space $\mathbb{C}^{\mathbb{Z}}$).

(c) Consider $u = (\ldots, 5, 0, 2, 3, 0, 0, \ldots)$ with period 6. Compute $(U(\omega))_{\omega=0}^{5}$ by writing $u = a + b$ for even a and odd b and using the linearity of the Fourier transform, and compare your answer to Exercise 2.4(d).

2.11 Set up the formulae to compute vertex numbers $v_{n,a}^{(h)}(x)$ and corridor numbers $c_{n,a}^{(h)}$ for each case below.

(a) $h = 5$, $a = 1$, two-way moves.
(b) $h = 4$, $a = 2$, two-way moves.
(c) $h = 5$, $a = 1$, three-way moves.

Use your formulae to construct tables of vertex and corridor numbers. Compare your results for parts (a) and (b) to Exercises 1.1 and 1.2, respectively.

2.12 Use the Binomial Theorem to express the three-way transition operator $T_M^n = (I + R + L)^n$ in terms of two-way transitions $T = R + L$, and use your derivation to prove (2.32).

2.13 Let $(F_k)_{k \in \mathbb{N}_0} = (F(k))_{k \in \mathbb{N}_0}$ be the sequence of Fibonacci numbers. Set $F(k) = 0$ if $k < 0$. Prove that $(I + R)^n[F](k) = F(k + n)$ for all $k \geq n$. Use this fact to prove the following.

$$\sum_{k=0}^{n} \binom{n}{k} F_k = F_{2n}$$

2.14 It seems reasonable that $\lim_{m \to \infty} \frac{1}{4^m}\binom{2m}{m} = 0$. Establish this. (*Hint:* Look up Stirling's approximation.) Use this fact to prove the one-dimensional case of Pólya's famous "Drunkard's Walk" theorem [47]—namely, that with probability 1, an infinite random walk on the points of $\mathbb{Z}$ revisits the origin infinitely often.[9] (*Hint:* translate this problem into a question about corridors of the form $\{-m, \ldots, 0, 1, \ldots, m\}$.)

---

[9] According to Pólya's theorem, this is also true of lattice walks in dimension 2. That is, if the "drunkard" is walking wandering around the streets of an infinite city grid, the drunkard will almost surely return to the same bar infinitely often, so to speak. However, in higher dimensions, the drunkard becomes hopelessly lost; i.e., the walk will remain outside any given ball, with probability 1 (see also [21]). This result is not just an idle curiosity; there are also interesting connections between random walks and electrical engineering [19, 62].

2.15 The expression $\frac{\sin(ta)}{\sin(t)}$ from Eq. (2.36) is reminiscent of the **Dirichlet kernel** of Fourier analysis (e.g. see [21]). Suppose $a = 2j + 1$. Using the identity (see Exercise A.10),

$$\frac{\sin([2j+1]t)}{\sin(t)} = \sum_{k=-j}^{j} e^{2kti},$$

along with the integral formula for corridor numbers (2.37) derive the formula:

$$c_{2m,2j+1}^{(\infty)} = \sum_{k=-j}^{j} \binom{2m}{m-k}$$

(Hence, $c_{2m,2j+1}^{(\infty)}$ is equal to the sum of the $2j + 1$ binomial coefficients in the center of row $2m$ of Pascal's triangle. Moreover, when $a = 1$ (i.e. $j = 0$), we have $c_{2m}^{(\infty)} = \binom{2m}{m}$ as expected).

Then use this result to bootstrap the other three cases, $c_{2m,2j}^{(\infty)}$, $c_{2m+1,2j}^{(\infty)}$, and $c_{2m+1,2j}^{(\infty)}$, ultimately deriving the following.

$$c_{n,a}^{(\infty)} = \sum_{k=\lfloor -\frac{a-1}{2} \rfloor}^{\lfloor \frac{a-1}{2} \rfloor} \binom{n}{\lfloor \frac{n}{2} \rfloor - k} \tag{2.43}$$

2.16 Interpreting (2.31) as a Riemann sum, derive the following formula for the number of infinite three-way corridor paths with initial point $a = 2j$, for $j \geq 1$.

$$M_{n,2j}^{(\infty)} = \frac{1}{2} \sum_{k=0}^{\lceil n/2 \rceil} \left[\binom{n}{2k} + \binom{n+1}{2k} \right] \sum_{\ell=-j}^{j} \binom{2k}{k+\ell} \tag{2.44}$$

Then work out the formulae for odd $a \in \mathbb{N}$.

Research Questions

2.17 As mentioned in Research Question 1.18, the Steck matrix can be used to count lattice paths in great generality, however the computations can be time-consuming, especially as the path length increases.[10] Consider the converse question: What can we learn about certain matrix determinants based on their connections to vertex numbers via Formula (2.23)?

---

[10]Using a fast matrix multiplication algorithm, the time complexity for finding the determinant of an $n \times n$ matrix is roughly $O(n^{2.373})$ [2].

2.18 Study the abstract Fourier theory for an orthonormal basis $\mathbf{u}_1, \mathbf{u}_2, \ldots, \mathbf{u}_n$ of $\mathbb{C}^n$ with respect to an inner product $\langle \mathbf{u}, \mathbf{v} \rangle$ (see e.g. [21, 55]). In particular, develop Parseval's Theorem in this general setting. How does our finite Fourier transform fit into this setting? Although we only use general properties of Fourier transforms, report on the importance of the Cooley-Tukey algorithm.

Chapter 3
Lattice Walks in Higher Dimensions

We now explore the possibility of counting paths in higher dimensional analogs of corridors, using more general types of transition operators. Then Theorem 2.1 and Formula 2.29 become special cases of a more general construction. We are certainly not the first to consider higher dimensional lattice paths (see e.g., [30, 36]), however our methods provide an alternative to the more traditional methods while at the same time allowing for paths and walks in which the steps are not confined just to directions parallel to the axes. Moreover, our Fourier techniques lead to further results and generalizations to be developed in Chap. 4 and Appendix B.

3.1 Multidimensional Corridors

We begin by generalizing the definitions of the previous chapters and introducing a few new terms that will aid in our discussion.

Definition 3.1 Let $r \in \mathbb{N}$ represent the dimension.

- A **path function** is a function $\mathbf{p} : \mathbb{N}_0 \to \mathbb{Z}^r$ denoting the position of a lattice path at each time (or step) $n \in \mathbb{N}_0$.
- A **move** is an element $\mathbf{m} \in \{-1, 0, 1\}^r$ representing a single step along a path, via vector addition: $\mathbf{p}(k+1) = \mathbf{p}(k) + \mathbf{m}$. The move $\mathbf{m} = (0, 0, \ldots, 0)$ is called the *identity move* and corresponds to a step that does not change position.
- A **path** of length n corresponding to a path function $\mathbf{p}$ is the lattice path with vertices $(k, \mathbf{p}(k))$ for $k = 0, 1, \ldots, n$.

Moves in one and two dimensions are illustrated in Fig. 3.1. In any given situation, the moves will be chosen from a prescribed set of *allowable moves*, $\mathscr{M} = \{\mathbf{m}_1, \mathbf{m}_2, \ldots, \mathbf{m}_N\}$, a subset of all possible moves for a given dimension, per-

© Springer Nature Switzerland AG 2019
S. Ault and C. Kicey, *Counting Lattice Paths Using Fourier Methods*, Applied and Numerical Harmonic Analysis,
https://doi.org/10.1007/978-3-030-26696-7_3

Fig. 3.1 There are 2 nonidentity moves in $\mathbb{Z}^1$ and 8 nonidentity moves in $\mathbb{Z}^2$

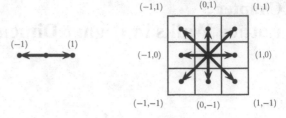

haps including the identity move. However as we shall see, there are certain restrictions on the set $\mathcal{M}$ that are required in order for the upcoming Fourier methods to work; the set $\mathcal{M}$ must be *balanced*—more on this point later.

Example 3.1 One-dimensional corridor paths have a set of two allowable moves, $\mathcal{M} = \{(1), (-1)\}$. The path shown in Fig. 1.2 (in Sect. 1.1) corresponds to a path function p with $p(k)$ taking on the values $2, 3, 2, 3, 2, 1, 2, 3$ for $k \in \{0, \ldots, 7\}$. Three-way corridor paths also include the identity move, $\mathcal{M} = \{(1), (0), (-1)\}$. The path shown in Fig. 1.6 (in Sect. 1.3) corresponds to a path function p with $p(k)$ taking on the values $3, 3, 2, 3, 4, 4, 3, 3, 2, 1$ for $k \in \{0, \ldots, 9\}$.

Our paths will generally be constrained by rectangular bounds, defining the multidimensional analog of a corridor, lattice points in an r-dimensional box.

Definition 3.2 Fix $h_j \in \mathbb{N}$ for $j = 1, 2, \ldots, r$, and let $\mathbf{h} = (h_1, \ldots, h_r)$. The **fundamental region** $\mathcal{C}^\mathbf{h}$ is the following set of lattice points.

$$\mathcal{C}^\mathbf{h} = \prod_{j=1}^{r} \{1, 2, \ldots, h_j\} = \{(x_1, \ldots, x_r) \in \mathbb{Z}^r \mid 1 \leq x_j \leq h_j, \forall j\}$$

The **h-corridor** is the set:

$$\mathbb{N}_0 \times \mathcal{C}^\mathbf{h} = \{(n, x_1, \ldots, x_r) \in \mathbb{N}_0 \times \mathbb{Z}^r \mid 1 \leq x_j \leq h_j, \forall j\}$$

See Fig. 3.2 for an illustration of a fundamental region in the case $r = 2$.

Definition 3.3 A **corridor path** of length n in the **h**-corridor, with respect to a fixed set of allowable moves $\mathcal{M}$ and chosen initial point $\mathbf{a} \in \mathcal{C}^\mathbf{h}$, is a path whose path function $\mathbf{p}$ satisfies the following rules:

1. $\mathbf{p}(0) = \mathbf{a}$.
2. $\mathbf{p}(k) \in \mathcal{C}^\mathbf{h}, \forall 0 \leq k \leq n$,
3. $\mathbf{p}(k+1) = \mathbf{p}(k) + \mathbf{m}$ for some $\mathbf{m} \in \mathcal{M}, \forall 0 \leq k < n$.

If the set of allowable moves $\mathcal{M}$ does not contain the identity move, then a corridor path in the lattice $\mathbb{N}_0 \times \mathcal{C}^\mathbf{h}$ may be interpreted as a walk in a graph $G_{\mathbf{h}, \mathcal{M}}$ on the vertices of $\mathcal{C}^\mathbf{h}$ whose edges correspond to the allowable moves in $\mathcal{M}$ (analogous to how we defined walks in Chap. 1). Again, only certain move sets $\mathcal{M}$ make sense

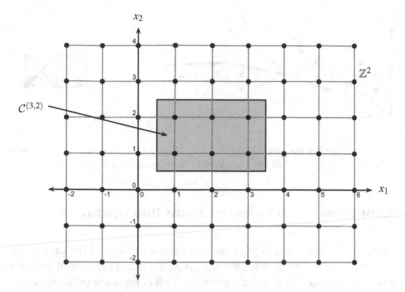

Fig. 3.2 The fundamental region $C^{(3,2)}$ defining the $(3, 2)$-corridor

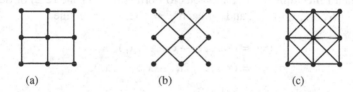

(a)   (b)   (c)

Fig. 3.3 Graphs $G_{\mathbf{h},\mathscr{M}}$ for $\mathbf{h} = (3, 3)$ and various move sets $\mathscr{M}$: **a** $\mathscr{M} = \{(\pm 1, 0), (0, \pm 1)\}$; **b** $\mathscr{M} = \{(1, 1), (1, -1), (-1, 1), (-1, -1)\}$; (c) $\mathscr{M} =$ all nonidentity moves

under this interpretation. Figure 3.3 shows a few graphs corresponding to different move sets when $r = 2$.

Example 3.2 Figure 3.4 shows a path in the $(3, 2)$-corridor in which all moves are allowed, along with the interpretation of that path as a walk in a graph.

Definition 3.4 Given a corridor with specified initial position $\mathbf{x} = \mathbf{a}$ and set of allowable moves, for each $n \in \mathbb{N}_0$, the **state function** is a function $v_n : \mathbb{Z}^r \to \mathbb{C}$ such that whenever $\mathbf{x} \in C^{\mathbf{h}}$, $v_n(\mathbf{x})$ equals the number of n-length corridor paths whose path function $\mathbf{p}$ has $\mathbf{p}(0) = \mathbf{a}$ and $\mathbf{p}(n) = \mathbf{x}$.

Observe that Definition 3.4 only specifies the values of state functions on $\mathbf{x} \in C^{\mathbf{h}}$. In order for us to use our Fourier methods, we will need v_n to have domain in $\mathbb{Z}^r$ and satisfy further properties. We need a multidimensional analog to the concept of *admissibility* from Sect. 2.1.

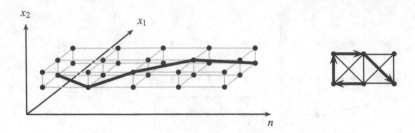

Fig. 3.4 *(Left)* A path in the corridor $\mathbb{N}_0 \times \mathcal{C}^{(3,2)}$. *(Right)* Corresponding walk in the graph $G_{(3,2),\mathcal{M}}$, where $\mathcal{M}$ is the set of all eight nonidentity two-dimensional moves

3.2   Admissibility and Balanced Transition Operators

Just as in Chap. 2, it becomes convenient to define $d_j = h_j + 1$ for each j; that is, $\mathbf{d} = \mathbf{h} + 1$ or $\mathbf{h} = \mathbf{d} - 1$ (where we use the convention that the sum of a vector and a scalar will mean to add that scalar amount to each component of the vector).

If $\mathbf{d} = (d_1, d_2, \ldots, d_r)$ is given (with $d_j \in \mathbb{N}$ for each $j = 1, 2, \ldots, r$), we may define certain **reflections** of $\mathbb{Z}^r$, analogous to Formulae (2.1) and (2.2) of Sect. 2.1. For all $\mathbf{x} = (x_1, \ldots, x_r) \in \mathbb{Z}^r$, and for each $j = 1, 2, \ldots, r$, define:

$$\rho_j(\mathbf{x}) = (x_1, \ldots, -x_j, \ldots, x_r), \tag{3.1}$$

$$\sigma_j(\mathbf{x}) = (x_1, \ldots, 2d_j - x_j, \ldots, x_r), \tag{3.2}$$

$$\tau_j = \sigma_j \rho_j \tag{3.3}$$

Note that τ_j is translation by $2d_j$ in the x_j component. Admissibility is defined in terms of these reflections.

Definition 3.5 A function $v : \mathbb{Z}^r \to \mathbb{C}$ is called **admissible** if v satisfies:

$$v\rho_j = -v \quad \text{for each } j, \text{ and} \tag{3.4}$$

$$v\sigma_j = -v \quad \text{for each } j \tag{3.5}$$

As in the one-dimensional case, we find that an admissible state function satisfies certain antisymmetry and periodicity properties, but now with respect to each coordinate:

1. Antisymmetry with respect to the hyperplane $x_j = 0$.
2. Antisymmetry with respect to the hyperplane $x_j = d_j$.
3. Periodicity in the x_j component with period $2d_j$.

It is worth noting that these three properties force $v(\mathbf{x}) = 0$ whenever the jth component of $\mathbf{x}$ is a multiple of d_j (for any j). Defining admissible states in this way is equivalent to our original notion of a *dual corridor structure*. Observe, if v_n is

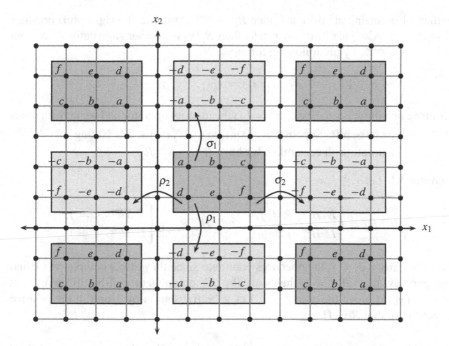

Fig. 3.5 Dual corridor structure for $\mathcal{C}^{(3,2)}$. Reflections ρ_1, ρ_2, σ_1, and σ_2 flip and negate the values as shown. The state has value zero at all points not otherwise labeled

admissible, then the values of v_n are determined by its values in $\mathcal{C}^{\mathbf{d}-1}$ (see Fig. 3.5 for example).

We now introduce a family of linear operators that acts on the states and serves to encode the allowable moves. If γ is any invertible transformation of $\mathbb{Z}^r$ (such as a reflection or translation), then define an associated operator H^γ that acts on functions $v : \mathbb{Z}^r \to \mathbb{C}$ by:

$$H^\gamma[v](\mathbf{x}) = v(\gamma^{-1}\mathbf{x}), \quad \forall \mathbf{x} \in \mathbb{Z}^r \tag{3.6}$$

A more concise way to state (3.6) is to say that: $H^\gamma[v] = v\gamma^{-1}$ for every function v. If η, γ are two transformations, then the composition satisfies $H^\eta H^\gamma = H^{\eta\gamma}$ (see Exercise 3.3). If γ is a reflection, then we have $\gamma^{-1} = \gamma$, so we could have defined admissible states as those for which $H^{\rho_j}[v] = -v$ and $H^{\sigma_j}[v] = -v$ for each j. We are also interested in **shift** (or **translation**) operators. Suppose that $\mathbf{m} \in \mathbb{Z}^r$ is any vector; if $\gamma_\mathbf{m}$ represents translation of points in $\mathbb{Z}^r$ by $\mathbf{m}$, then the following holds:

$$H^{\gamma_\mathbf{m}}[v](\mathbf{x}) = v(\gamma_\mathbf{m}^{-1}\mathbf{x}) = v(\mathbf{x} - \mathbf{m}) \tag{3.7}$$

For convenience, we use the notation $R^\mathbf{m}$ in place of $H^{\gamma_\mathbf{m}}$. Observe, if

$$\mathbf{m} = (0, \ldots, 0, 1, 0, \ldots, 0),$$

with the 1 occurring in position j, then $R_j = R^{\mathbf{m}}$ represents the **right-shift** operator along the x_j-axis. That is, if v is a state, then $R_j[v]$ is another state defined on input $(x_1, \ldots, x_r) \in \mathbb{Z}^r$ by the following formula.

$$R_j[v](x_1, \ldots, x_j, \ldots, x_r) = v(x_1, \ldots, x_j - 1, \ldots, x_r)$$

As in the one-dimensional case, $R_j^0 = I$ is the identity operator and we may interpret $R_j^{-1} = L_j$ as **left-shift**. Suppose $\mathbf{m} = (m_1, m_2, \ldots, m_r) \in \mathbb{Z}^r$. Noting that the R_j operators commute with each other, we have $R^{\mathbf{m}} = R_1^{m_1} R_2^{m_2} \cdots R_r^{m_r}$.

Lemma 3.1 *Let* $1 \le j, k \le r$.

$$H^{\rho_k} R_j = \begin{cases} R_j H^{\rho_k} & k \neq j, \\ L_j H^{\rho_k} & k = j \end{cases} \quad and \quad H^{\sigma_k} R_j = \begin{cases} R_j H^{\sigma_k} & k \neq j, \\ L_j H^{\sigma_k} & k = j \end{cases}$$

Proof Clearly if $k \neq j$, the operators commute since they affect components k and j separately. Now if $k = j$, then only the jth component will be affected. So it is sufficient to prove the case $r = 1$. Let γ be translation by 1 unit in the positive direction, so that $R = H^\gamma$.

$$\rho\gamma(x) = -(x - 1) = -x + 1 = \gamma^{-1}\rho(x)$$

The above shows that $H^\rho R = H^\rho H^\gamma = H^{\rho\gamma} = H^{\gamma^{-1}\rho} = R^{-1} H^\rho = LH^\rho$. Similarly,

$$\sigma\gamma(x) = 2d - (x - 1) = 2d - x + 1 = \gamma^{-1}\sigma(x),$$

which proves that $H^\sigma R = R^{-1} H^\sigma = LH^\sigma$. $\square$

Definition 3.6 The **transition** operator corresponding to a set of moves $\mathcal{M}$ is defined by the following:

$$T = T_{\mathcal{M}} = \sum_{\mathbf{m} \in \mathcal{M}} R^{\mathbf{m}} \tag{3.8}$$

As in the one-dimensional case, the transition operator serves to encode the allowable moves for our paths. Suppose for the moment that there are no restrictions at all (no corridor) and we would like to count the number of paths in $\mathbb{N}_0 \times \mathbb{Z}^r$ beginning at the origin and ending at $\mathbf{x} \in \mathbb{Z}^r$, but allowing only the moves from some specified move set $\mathcal{M}$. If we have a state function $v_n(\mathbf{x})$ that counts all such paths of length n and ending at $\mathbf{x}$, then

$$T[v_n](\mathbf{x}) = \sum_{\mathbf{m} \in \mathcal{M}} R^{\mathbf{m}}[v_n](\mathbf{x}) = \sum_{\mathbf{m} \in \mathcal{M}} v_n(\mathbf{x} - \mathbf{m})$$

would count precisely the number of $(n + 1)$-length paths ending at $\mathbf{x}$. In other words, $T[v_n] = v_{n+1}$.

For corridor numbers, on the other hand, we must somehow restrict moves at the boundaries of the corridor. However, instead of restricting moves at the boundary, our methods keep the transition function intact and instead make use of properties of the state functions themselves.

Recall, if a state v is admissible (Definition 3.5), then $v(\mathbf{x}) = 0$ whenever $x_j = 0$ or $x_j = d_j$. That is, v has a *buffer zone* of zeros, one unit in thickness, separating the points of the corridor from the other points of $\mathbb{Z}^r$. This buffer zone permits $T[v_n]$ to correctly count corridor numbers v_{n+1}, even at the boundaries, because the zero values in the buffer zone do not contribute to the vertex numbers. The key now is to use only transition operations that preserve admissibility of states. Such transition operators will be called *balanced*.

In the following definition, note that a reflection ρ_j may also be applied to a move **m** to obtain a new move $\rho_j(\mathbf{m})$. For example, $\rho_2((1, 1, -1)) = (1, -1, -1)$.

Definition 3.7

- A set of moves $\mathscr{M} = \{\mathbf{m}_1, \ldots, \mathbf{m}_N\}$ is called **balanced** if $\mathscr{M}$ is invariant under each of the reflection operators ρ_j for $j = 1, 2, \ldots, r$. That is, for every j, we have $\{\rho_j(\mathbf{m}_1), \ldots, \rho_j(\mathbf{m}_N)\} = \mathscr{M}$.
- A transition operator $T = T_{\mathscr{M}}$ is called **balanced** if $\mathscr{M}$ is a balanced set of moves.

As a consequence of Definition 3.7, $\mathscr{M}$ is balanced if whenever there is a move $\mathbf{m} = (m_1, \ldots, m_r) \in \mathscr{M}$, then so is $(\pm m_1, \ldots, \pm m_r) \in \mathscr{M}$ for all the possible ways of choosing the sign of each "$\pm$." Equivalently, $\mathscr{M}$ is balanced if the graph $G_{\mathbf{h}, \mathscr{M}}$ is symmetric with respect to reflection across every coordinate hyperplane $x_j = 0$.

Example 3.3 When $r = 1$, there are only three nonzero balanced transition operators: I ($\mathscr{M} = \{(0)\}$), $R_1 + R_1^{-1}$ ($\mathscr{M} = \{(1), (-1)\}$), and $R_1 + I + R_1^{-1}$ ($\mathscr{M} = \{(1), (0), (-1)\}$), which correspond, respectively, to the identity transition, the two-way corridor path transition, and the three-way corridor path transition.

Equivalently, a transition operator T is balanced if and only if $H^{\rho_j}T = T$ for every $j \in \{1, 2, \ldots, r\}$. Define for each j, an operator $T_j = R_j + R_j^{-1} = R_j + L_j$.

Lemma 3.2 *For any* $j, k \in \{1, 2, \ldots, r\}$, *we have* $H^{\rho_k}T_j = T_j H^{\rho_k}$ *and* $H^{\sigma_k}T_j = T_j H^{\sigma_k}$.

The proof, which relies on Lemma 3.1, is left as an easy exercise for the reader.

Lemma 3.3 *A transition operator T is balanced if and only if T is a sum of square-free products of T_j for various $j \in \{1, 2, \ldots, r\}$.*

Proof Consider a balanced transition operator T as in (3.8). The terms of T must necessarily be square-free since the components of each move are either 0 or ± 1. Since T is balanced, we have $(m_1, \ldots, m_{r-1}, 1) \in \mathscr{M}$ if and only if $(m_1, \ldots, m_{r-1}, -1) \in \mathscr{M}$. Thus, $T_r = R_r + L_r$ can be factored from each such pair, and then we may write $T = U + VT_r$, where U and V are operators involving only the shifts R_j and L_j for $j \in \{1, 2, \ldots, r - 1\}$. The result follows by recursively factoring U and V in the same way.

The reverse direction follows immediately from Lemma 3.2.   □

Fig. 3.6 Left: an unbalanced
transition, $R_1R_2 + L_1L_2$.
Right: a balanced transition,
$R_1R_2 + R_1L_2 + L_1R_2 +$
L_1L_2

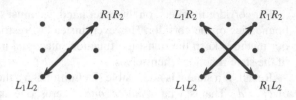

Example 3.4 $T = R_1R_2 + L_1L_2$ is not balanced, because it is not reflection-invariant.

$$H^{\rho_1}T = H^{\rho_1}R_1R_2 + H^{\rho_1}L_1L_2 = L_1H^{\rho_1}R_2 + R_1H^{\rho_1}L_2 = (L_1R_2 + R_1L_2)H^{\rho_1}$$

On the other hand, $S = R_1R_2 + R_1L_2 + L_1R_2 + L_1L_2$ is balanced. In fact, S can be factored as follows.

$$S = R_1R_2 + R_1L_2 + L_1R_2 + L_1L_2 = (R_1 + L_1)(R_2 + L_2) = T_1T_2$$

Figure 3.6 illustrates both of these examples.

Henceforth, we will only consider balanced transition operators. Furthermore, we will focus on state functions defined inductively by $v_{n+1} = T[v_n]$, where an initial state v_0 is given, so that in general,

$$v_n = T^n[v_0], \text{ for } n \in \mathbb{N}_0 \tag{3.9}$$

The key to showing that Eq. (3.9) serves to count the vertex numbers for a corridor is to prove that T takes admissible states to admissible states.

Lemma 3.4 *Suppose T is a balanced transition operator. If v is admissible, then so is $T[v]$.*

Proof By Lemma 3.3, it suffices to show that T_j preserves admissibility for an arbitrary $j \in \{1, 2, \ldots, r\}$. We must show that $H^{\rho_k}\left[T_j[v]\right] = -T_j[v]$ and $H^{\sigma_k}\left[T_j[v]\right] = -T_j[v]$. By Lemmas 3.3 and 3.2, and since v is admissible, we have the following results for $\gamma = \rho_k$ and $\gamma = \sigma_k$.

$$H^{\gamma}\left[T_j[v]\right] = (H^{\gamma}T_j)[v] = (T_jH^{\gamma})[v] = T_j\left[H^{\gamma}[v]\right] = T_j[-v] = -T_j[v]$$

Hence $T[v]$ is also admissible. □

Theorem 3.1 *Suppose T is a transition operator corresponding to a balanced set of moves $\mathcal{M}$. The number of corridor paths of length n with initial point $\mathbf{a} \in C^{\mathbf{h}}$ and ending at position $\mathbf{x} \in C^{\mathbf{h}}$, allowing only the moves in $\mathcal{M}$, is equal to $v_n(\mathbf{x}) = T^n[v_0](\mathbf{x})$, where the initial state $v_0(\mathbf{x})$ is the admissible state function such that $v_0(\mathbf{a}) = 1$ and $v_0(\mathbf{x}) = 0$ for $\mathbf{x} \in C^{\mathbf{h}}, \mathbf{x} \neq \mathbf{a}$.*

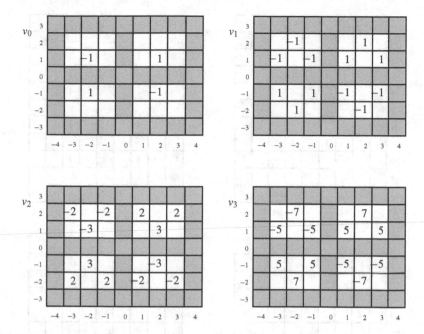

Fig. 3.7 Example corridor state progression $v_0 \to v_1 \to v_2 \to v_3$ with $\mathbf{a} = (2, 1)$, $\mathbf{d} = (4, 3)$, and $T = T_1 + T_2$

Proof By convention there is exactly 1 path of length 0, the one that begins and ends at the initial point, so $v_0(\mathbf{x})$ indeed counts the 0-length paths from $\mathbf{a}$ to $\mathbf{x}$. Now if $v_{n-1}(\mathbf{x})$ is an admissible state function that counts the paths of length $n - 1$ ending at $\mathbf{x}$, then $T[v_{n-1}]$ is admissible by Lemma 3.4 and induction. Thus, $v_n = T[v_{n-1}]$ counts the n-length paths as required. $\qquad\square$

Example 3.5 Figure 3.7 shows the dual corridor structure for $\mathcal{C}^{(3,2)}$ ($\mathbf{d} = \mathbf{h} + 1 = (4, 3)$) and transition function $T = R_1 + L_1 + R_2 + L_2 = T_1 + T_2$. Note that the initial state v_0 corresponds to a path that begins at point $(2, 1)$. Only the nonzero values of v_n are indicated.

A diagonal transition operator, $T = R_1 R_2 + R_1 L_2 + L_2 R_1 + L_2 L_2 = T_1 T_2$, is illustrated in Fig. 3.8 with initial point $(1, 2)$ in $\mathcal{C}^{(3,4)}$.

3.3 Fourier Analysis of Multidimensional Corridors

Our main purpose in this chapter is to count the number of n-length walks within a bounded rectangular region of $\mathbb{Z}^r$ with specified starting and ending points and with moves from a given balanced move set. We shall see that this very general problem

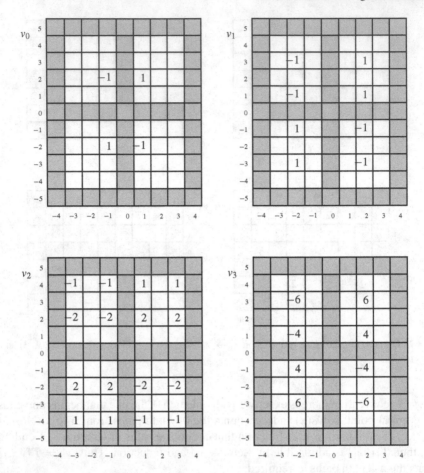

Fig. 3.8 Example corridor state progression $v_0 \to v_1 \to v_2 \to v_3$ with $\mathbf{a} = (1, 2)$, $\mathbf{d} = (4, 5)$, and $T = T_1 T_2$

can be addressed using a single tool: the **multidimensional DFT** (discrete Fourier transform).

Fourier transforms may be defined on a variety of function and sequence spaces, in which a function or *signal* is analyzed using periodic functions with various frequencies. As discussed previously for the one-dimensional case in Chap. 2, the Fourier transform is useful to this kind of combinatorial problem because it acts like a periodic generating function, serving to encode our periodic dual corridor structure. Moreover the shift operators such as those making up the transition operators as in Definition 3.6 correspond to multiplication of functions on the frequency side. We focus on what is necessary for our purposes, a variation of the classic discrete Fourier transform in arbitrary dimension. This material is standard and can be found in [21, 55] for instance; however, we do not require extensive theoretical knowledge of Fourier analysis.

Suppose u is a function of discrete variables $\mathbf{x} = (x_1, x_2, \ldots, x_r) \in \mathbb{Z}^r$, with periodicity $N_j \geq 1$ in the jth component. Let $\omega = (\omega_1, \omega_2, \ldots, \omega_r) \in \mathbb{Z}^r$ and $\mathbf{N} = (N_1, N_2, \ldots, N_r)$. Then the **DFT** of u is a (complex-valued) function of ω given by

$$\mathcal{F}[u](\omega) = U(\omega) = \sum_{\mathbf{x}=0}^{\mathbf{N}-1} u(\mathbf{x}) e^{-2\pi i \omega \cdot (\mathbf{x}/\mathbf{N})}, \tag{3.10}$$

where $\mathbf{N} - 1 = (N_1 - 1, N_2 - 1, \ldots, N_r - 1), \mathbf{x}/\mathbf{N} = (x_1/N_1, x_2/N_2, \ldots, x_r/N_r)$, and the sum is understood to be over all $\mathbf{x}$ such that $0 \leq x_j \leq N_j - 1$ in each component. Similarly, the **inverse DFT** of a function $U(\omega)$ is a function of $\mathbf{x}$ given by the formula:

$$\mathcal{F}^{-1}[U](\mathbf{x}) = u(\mathbf{x}) = \frac{1}{\prod_{j=1}^{r} N_j} \sum_{\omega=0}^{\mathbf{N}-1} U(\omega) e^{2\pi i \mathbf{x} \cdot (\omega/\mathbf{N})} \tag{3.11}$$

It is straightforward to show that $\mathcal{F}^{-1}[\mathcal{F}[u]] = u$ (in other words, $\mathcal{F}^{-1}\mathcal{F} = I$ as operators). There are no issues of convergence since the sums are finite.

Let's explore how $\mathcal{F}$ works on a function of particular importance. Consider the **delta function**, which we define on $\mathbb{Z}^r$ as follows:

$$\delta(\mathbf{x}) = \begin{cases} 1 & \mathbf{x} = \mathbf{0}, \\ 0 & \mathbf{x} \neq \mathbf{0} \end{cases}$$

The function δ has value zero at all points except the origin at which its value is 1. Using a shift operator, we can create a delta function $\delta_{\mathbf{a}}$ whose only nonzero value occurs at $\mathbf{a} \in \mathbb{Z}^r$.

$$\delta_{\mathbf{a}}(\mathbf{x}) = R^{\mathbf{a}}[\delta](\mathbf{x}) = \delta(\mathbf{x} - \mathbf{a})$$

Let $\mathbf{N} = (N_1, \ldots, N_r) \in \mathbb{N}^r$, and suppose $\mathbf{a} = (a_1, \ldots, a_r) \in \mathbb{Z}^r$ such that $0 \leq a_j < N_j$ for each j. Observe that $\delta_{\mathbf{a}}$ is not periodic, but because the DFT is defined as a sum over $\mathbf{x}$ in a single periodic block, we may find $\mathcal{F}[\delta_{\mathbf{a}}]$ as if it were $\mathbf{N}$-periodic.

$$\begin{aligned} \mathcal{F}[\delta_{\mathbf{a}}](\omega) &= \sum_{\mathbf{x}=0}^{\mathbf{N}-1} \delta_{\mathbf{a}}(\mathbf{x}) e^{-2\pi i \omega \cdot (\mathbf{x}/\mathbf{N})} \\ &= e^{-2\pi i \omega \cdot (\mathbf{a}/\mathbf{N})} \\ &= \exp\left[-2\pi i \left(\frac{\omega_1 a_1}{N_1} + \cdots + \frac{\omega_r a_r}{N_r}\right)\right], \end{aligned} \tag{3.12}$$

where we have used $\exp(x) = e^x$ for better readability in the last line. Formula (3.12) becomes especially useful in the next section, where we will also make use of the property,

$$\delta_{(a_1,\dots,a_j,\dots,a_r)}\rho_j = \delta_{(a_1,\dots,-a_j,\dots,a_r)}, \tag{3.13}$$

as well as the following formula involving the shift operators.

$$\mathcal{F}\left[R_j^k[u]\right] = e^{-2k\pi i\omega_j/N_j}\mathcal{F}[u] \tag{3.14}$$

You will prove (3.13) in the exercises. Formula (3.14) follows from the one-dimensional case (recall Sect. 2.2).

Let $r \in \mathbb{N}$, $\mathbf{h} \in \mathbb{N}^r$, and set $\mathbf{d} = \mathbf{h} + 1$. We are now in good position to use the multidimensional DFT to count paths in the $\mathbf{h}$-corridor. The starting point and key result is Theorem 3.1. Through a series of lemmas, we will develop explicit formulae for the vertex state functions of higher dimensional corridors in Theorem 3.2.

Suppose v_n is an admissible state. In particular, v_n is $2\mathbf{d}$-periodic. So with $N_j = 2d_j$ in Eq. (3.10), the DFT V_n corresponding to v_n (re-indexed for convenience of calculations) is given by:

$$V_n(\omega) = \sum_{\mathbf{x}=-\mathbf{d}+1}^{\mathbf{d}} v_n(\mathbf{x})e^{-\pi i\omega\cdot(\mathbf{x}/\mathbf{d})} \tag{3.15}$$

We can recover v_n via the inverse DFT (3.11).

$$v_n(\mathbf{x}) = \frac{1}{2^r \prod_{j=1}^r d_j} \sum_{\omega=-\mathbf{d}+1}^{\mathbf{d}} V_n(\omega)e^{\pi i\mathbf{x}\cdot(\omega/\mathbf{d})} \tag{3.16}$$

Let $\mathbf{a} \in \mathcal{C}^{\mathbf{h}}$ be the initial point for our corridor paths. Then the initial state function satisfies $v_0(\mathbf{x}) = \delta_{\mathbf{a}}(\mathbf{x})$ for all $\mathbf{x} \in \mathcal{C}^{\mathbf{h}}$. However in order to ensure v_0 is admissible, it must satisfy the antisymmetry conditions with respect to ρ_j and σ_j for all $j \in \{1, \dots, r\}$. It is equivalent to make v_0 antisymmetric with respect to each coordinate plane and then extend by $2\mathbf{d}$-periodicity (see Exercise 2.3 in the one-dimensional case). To this end, let $\mathbf{k} = (k_1, \dots, k_r)$ and define

$$\Delta_{\mathbf{a}} = \sum_{\mathbf{k}\in\{0,1\}^r} (-1)^{k_1+k_2+\dots+k_r}\delta_{\mathbf{a}}\rho_1^{k_1}\rho_2^{k_2}\cdots\rho_r^{k_r}, \tag{3.17}$$

or equivalently, using Eq. (3.13), we have:

$$\Delta_{\mathbf{a}} = \sum_{\mathbf{k}\in\{0,1\}^r} (-1)^{\sum k_j}\delta_{((-1)^{k_1}a_1,\, (-1)^{k_2}a_2,\, \dots,\, (-1)^{k_r}a_r)} \tag{3.18}$$

As above, explicit periodization is not needed for the purpose of applying the DFT.

Example 3.6 Consider the corridor and initial state v_0 as in Example 3.5 and Fig. 3.7. All corridor paths begin at $(a_1, a_2) = (2, 1)$, so $v_0(2, 1) = 1$ and has value 0 elsewhere in the fundamental region. That is, $v_0(\mathbf{x}) = \delta_{(2,1)}(\mathbf{x})$ when $\mathbf{x} \in C^{(3,2)}$. Antisymmetrization of $\delta_{(2,1)}$ using (3.18), yields:

$$
\begin{aligned}
\Delta_{(2,1)} &= \sum_{(k_1,k_2)\in\{0,1\}^2} (-1)^{k_1+k_2} \delta_{((-1)^{k_1}\cdot 2,\, (-1)^{k_2}\cdot 1)} \\
&= (-1)^{0+0}\delta_{(+2,+1)} + (-1)^{0+1}\delta_{(+2,-1)} \\
&\quad + (-1)^{1+0}\delta_{(-2,+1)} + (-1)^{1+1}\delta_{(-2,-1)} \\
&= \delta_{(2,1)} - \delta_{(2,-1)} - \delta_{(-2,1)} + \delta_{(-2,-1)}
\end{aligned}
$$

Remark 3.1 Observe that the above implies that $\Delta_{(2,1)}(x_1, x_2) = \Delta_2(x_1)\Delta_1(x_2)$.

Lemma 3.5 *Let v_0 be an admissible initial state corresponding to the initial point $\mathbf{a} \in C^{d-1}$. The DFT of v_0 is given by:*

$$
V_0(\omega) = (-2i)^r \prod_{j=1}^{r} \sin\left(\frac{\pi \omega_j a_j}{d_j}\right) \tag{3.19}
$$

Proof Since v_0 is assumed to be admissible, v_0 is the periodization of $\Delta_{\mathbf{a}}$. Use (3.12) with Eq. (3.18), keeping in mind that v_0 has periodicity $N_j = 2d_j$ in the jth component.

$$
\begin{aligned}
V_0(\omega) &= \mathcal{F}[\Delta_{\mathbf{a}}](\omega) \\
&= \sum_{\mathbf{k}\in\{0,1\}^r} (-1)^{\sum k_j} \mathcal{F}\left[\delta_{((-1)^{k_1}a_1,\, (-1)^{k_2}a_2,\, \ldots,\, (-1)^{k_r}a_r)}\right] \\
&= \sum_{\mathbf{k}\in\{0,1\}^r} (-1)^{\sum k_j} \exp\left[-2\pi i\left(\frac{\omega_1(-1)^{k_1}a_1}{2d_1} + \cdots + \frac{\omega_r(-1)^{k_r}a_r}{2d_r}\right)\right] \\
&= \sum_{\mathbf{k}\in\{0,1\}^r} (-1)^{\sum k_j} \exp\left(\frac{-\pi i\omega_1(-1)^{k_1}a_1}{d_1}\right) \cdots \cdot \exp\left(\frac{-\pi i\omega_r(-1)^{k_r}a_r}{d_r}\right)
\end{aligned}
$$

Let $E_j(k_j) = \exp\left(\frac{-\pi i\omega_j(-1)^{k_j}a_j}{d_j}\right)$, and note how the sum factors:

$$
\sum_{\mathbf{k}\in\{0,1\}^r} (-1)^{\sum k_j} \prod_{j=1}^{r} E_j(k_j) = \sum_{\mathbf{k}\in\{0,1\}^r} \prod_{j=1}^{r} (-1)^{k_j} E_j(k_j)
$$

$$
= \prod_{j=1}^{r} \sum_{k_j=0}^{1} (-1)^{k_j} E_j(k_j)
$$

Finally, use Euler's Formula on each factor.

$$\sum_{k_j=0}^{1} (-1)^{k_j} E_j(k_j) = \exp\left(\frac{-\pi i \omega_j a_j}{d_j}\right) - \exp\left(\frac{\pi i \omega_j a_j}{d_j}\right) = -2i \sin\left(\frac{\pi \omega_j a_j}{d_j}\right)$$

Then Eq. (3.19) follows.   □

Observe that $V_0(\omega)$ is real when r is even and purely imaginary when r is odd, a fact that will serve to simplify our results later on.

Remark 3.2 The above derivation could be streamlined by working out a type of *separation formula* for the $\Delta_{\mathbf{a}}$ function. See Exercises 3.7 and 3.8.

$$\Delta_{\mathbf{a}}(\mathbf{x}) = \Delta_{a_1}(x_1)\Delta_{a_2}(x_2)\cdots\Delta_{a_r}(x_r) \tag{3.20}$$

Example 3.7 With v_0 as in Example 3.5, we have:

$$V_0(\omega_1, \omega_2) = -4\sin\left(\frac{\pi\omega_1}{2}\right)\sin\left(\frac{\pi\omega_2}{3}\right)$$

Now recall from (3.9), the nth state function is determined by $v_n = T^n[v_0]$, for a transition operator T. In the following lemma, we show how the DFT of a state function changes after applying T.

Lemma 3.6 *Let v be an admissible state, and suppose T is the transition operator corresponding to a balanced set of moves $\mathscr{M}$. Then there is a function $\widehat{T}(\omega)$, that fits into the following equation.*

$$\mathcal{F}[T[v]] = \widehat{T} \cdot \mathcal{F}[v]$$

Moreover, let $\mathscr{M}^+ \subset \mathscr{M}$ be the subset of moves having only nonnegative entries; then we may write:

$$\widehat{T}(\omega) = \sum_{\mathbf{m}\in\mathscr{M}^+} \prod_{j=1}^{r} \left(2\cos\left(\frac{\pi\omega_j}{d_j}\right)\right)^{m_j} \tag{3.21}$$

Proof By Lemma 3.3 and linearity of $\mathcal{F}$, it is sufficient to consider $\mathcal{F}[T_j[v]]$ (see also Exercise 3.10). Making use of (3.14) with $N_j = 2d_j$, and after applying Euler's Formula, we obtain:

$$\mathcal{F}[T_j[v]] = \mathcal{F}[R_j[v]] + \mathcal{F}\left[R_j^{-1}[v]\right]$$
$$= e^{-\pi i \omega_j/d_j}\mathcal{F}[v] + e^{\pi i \omega_j/d_j}\mathcal{F}[v]$$
$$= 2\cos\left(\frac{\pi\omega_j}{d_j}\right)\mathcal{F}[v]$$

                                                                                            □

Example 3.8 The one-dimensional two-way transition operator $T = R + L$ has associated function $\widehat{T}(\omega) = 2\cos(\frac{\pi\omega}{d})$. The three-way transition operator $T_M = R + I + L$ corresponds with $\widehat{T_M}(\omega) = 1 + 2\cos(\frac{\pi\omega}{d})$. (Compare the corresponding formulae developed in Chap. 2.)

Example 3.9 Consider the $(2, 4, 3)$-corridor in dimension $r = 3$. That is, the fundamental region $\mathcal{C}^{(2,4,3)}$ is a sublattice of $\mathbb{N}^3$, and the corridor $\mathcal{C}^{(2,4,3)} \times \mathbb{N}_0$ exists within four-dimensional space. Note that $\mathbf{d} = (3, 5, 4)$ in this example.

Consider the following set of moves.

$$\mathcal{M} = \{(0, 0, 0), (1, 1, 0), (1, -1, 0), (-1, 1, 0), (-1, -1, 0), (1, 0, 1), (1, 0, -1),$$
$$(-1, 0, 1), (-1, 0, -1), (0, 1, 1), (0, 1, -1), (0, -1, 1), (0, -1, -1)\}$$

The corresponding transition function $T = T_{\mathcal{M}}$ is shown below.

$$T = I + R_1 R_2 + R_1 L_2 + L_1 R_2 + L_1 L_2$$
$$+ R_1 R_3 + R_1 L_3 + L_1 R_3 + L_1 L_3$$
$$+ R_2 R_3 + R_2 L_3 + L_2 R_3 + L_2 L_3$$

T is balanced because it can be factored to obtain $T = I + T_1 T_2 + T_1 T_3 + T_2 T_3$, a sum of square-free products of T_j operators. Observe:

$$\mathcal{M}^+ = \{(0, 0, 0), (1, 1, 0), (1, 0, 1), (0, 1, 1)\}$$

Hence, according to (3.21), the formula for $\widehat{T}$ in this case would be as follows:

$$\widehat{T}(\omega) = 1 + 4\cos\left(\frac{\pi\omega_1}{3}\right)\cos\left(\frac{\pi\omega_2}{5}\right) + 4\cos\left(\frac{\pi\omega_1}{3}\right)\cos\left(\frac{\pi\omega_3}{4}\right) + 4\cos\left(\frac{\pi\omega_2}{5}\right)\cos\left(\frac{\pi\omega_3}{4}\right)$$

Finally we state our general theorem concerning the multidimensional corridor numbers (compare Theorem 2.1).

Theorem 3.2 *Let $r \in \mathbb{N}$, $\mathbf{h} \in \mathbb{N}^r$, and set $\mathbf{d} = \mathbf{h} + 1$. Given an initial point $\mathbf{a} = (a_1, \dots, a_r) \in \mathcal{C}^{\mathbf{h}}$ and a balanced set of moves $\mathcal{M}$, the nth state for the $\mathbf{h}$-corridor is obtained by*

$$v_n = \mathcal{F}^{-1}\left[\widehat{T}^n \mathcal{F}[\Delta_{\mathbf{a}}]\right],$$

where $\widehat{T}$ is defined as in Lemma 3.6 for the transition operator $T = T_{\mathcal{M}}$ corresponding to the given move set, and $\Delta_{\mathbf{a}}$ is defined by Formula (3.18).

More explicitly,

$$
v_n(\mathbf{x}) = \begin{cases} \dfrac{(-1)^{\frac{r}{2}}}{\prod_{j=1}^{r} d_j} \displaystyle\sum_{\omega=-\mathbf{d}+1}^{\mathbf{d}} \cos(\pi\mathbf{x}\cdot(\omega/\mathbf{d})) \left[\widehat{T}(\omega)\right]^n \prod_{j=1}^{r} \sin\left(\dfrac{\pi\omega_j a_j}{d_j}\right) & \text{if } r \text{ is even,} \\[2em] \dfrac{(-1)^{\frac{r-1}{2}}}{\prod_{j=1}^{r} d_j} \displaystyle\sum_{\omega=-\mathbf{d}+1}^{\mathbf{d}} \sin(\pi\mathbf{x}\cdot(\omega/\mathbf{d})) \left[\widehat{T}(\omega)\right]^n \prod_{j=1}^{r} \sin\left(\dfrac{\pi\omega_j a_j}{d_j}\right) & \text{if } r \text{ is odd,} \end{cases}
$$

$$(3.22)$$

where $\widehat{T}$ is given by (3.21).

Proof Let $T = T_{\mathcal{M}}$ be the transition operator corresponding to the given move set. By Theorem 3.1, $v_n = T^n[v_0]$, where v_0 is the admissible initial state corresponding to the given initial point $\mathbf{a}$. Thus by Lemma 3.6 and induction, we obtain:

$$
V_n = \mathcal{F}[v_n] = \mathcal{F}\left[T^n[v_0]\right] = \widehat{T}^n \mathcal{F}[v_0] = \widehat{T}^n V_0
$$

Then by Lemma 3.5, we have:

$$
V_n(\omega) = \left[\widehat{T}(\omega)\right]^n V_0(\omega) = \left[\widehat{T}(\omega)\right]^n (-2i)^r \prod_{j=1}^{r} \sin\left(\frac{\pi\omega_j a_j}{d_j}\right) \tag{3.23}
$$

Since $\widehat{T}$ is real, V_n is real when r is even and purely imaginary when r is odd.

Suppose that r is even. Then $(-2i)^r = 2^r(-1)^{r/2}$, and as we apply the inverse DFT to V_n, we find:

$$
v_n = 2^r(-1)^{\frac{r}{2}} \mathcal{F}^{-1}\left[\left[\widehat{T}(\omega)\right]^n \prod_{j=1}^{r} \sin\left(\frac{\pi\omega_j a_j}{d_j}\right)\right]
$$

$$
v_n(\mathbf{x}) = 2^r(-1)^{\frac{r}{2}} \frac{1}{2^r \prod_{j=1}^{r} d_j} \sum_{\omega=-\mathbf{d}+1}^{\mathbf{d}} e^{\pi i \mathbf{x}\cdot(\omega/\mathbf{d})} \left[\widehat{T}(\omega)\right]^n \prod_{j=1}^{r} \sin\left(\frac{\pi\omega_j a_j}{d_j}\right)
$$

$$
= \frac{(-1)^{\frac{r}{2}}}{\prod_{j=1}^{r} d_j} \sum_{\omega=-\mathbf{d}+1}^{\mathbf{d}} e^{\pi i \mathbf{x}\cdot(\omega/\mathbf{d})} \left[\widehat{T}(\omega)\right]^n \prod_{j=1}^{r} \sin\left(\frac{\pi\omega_j a_j}{d_j}\right) \tag{3.24}
$$

By Euler's Formula (A.4), we have:

$$
e^{\pi i \mathbf{x}\cdot(\omega/\mathbf{d})} = \cos(\pi\mathbf{x}\cdot(\omega/\mathbf{d})) + i\sin(\pi\mathbf{x}\cdot(\omega/\mathbf{d})) \tag{3.25}
$$

Finally, because we know a priori that $v_n(\mathbf{x})$ is a real function when $\mathbf{x} \in \mathbb{Z}^r$ (in fact, integer valued), we can ignore the imaginary part of (3.25), allowing (3.24) to be expressed without complex numbers.

$$
v_n(\mathbf{x}) = \frac{(-1)^{\frac{r}{2}}}{\prod_{j=1}^{r} d_j} \sum_{\omega=-\mathbf{d}+1}^{\mathbf{d}} \cos(\pi\mathbf{x}\cdot(\omega/\mathbf{d})) \left[\widehat{T}(\omega)\right]^n \prod_{j=1}^{r} \sin\left(\frac{\pi\omega_j a_j}{d_j}\right)
$$

The argument is similar for r odd; see Exercise 3.12.   □

Now that we have Theorem 3.2 at our disposal, let's use it to work out a number of examples.

Example 3.10 With $\mathbf{d} = (4, 3)$, $\mathbf{a} = (2, 1)$, and $T = L_1 + R_1 + L_2 + R_2 = T_1 + T_2$ as in Example 3.5, Theorem 3.2 gives:

$$v_n(\mathbf{x}) = \frac{-1}{4 \cdot 3} \sum_{\omega_1 = -3}^{4} \sum_{\omega_2 = -2}^{3} \cos\left(\pi \left[\frac{x_1 \omega_1}{4} + \frac{x_2 \omega_2}{3}\right]\right) [\widehat{T}(\omega)]^n \sin\left(\frac{\pi \omega_j \cdot 2}{4}\right) \sin\left(\frac{\pi \omega_j \cdot 1}{3}\right)$$

The formula for $\widehat{T}$ is found using Eq. (3.21).

$$\widehat{T}(\omega) = 2\cos\left(\frac{\pi \omega_1}{4}\right) + 2\cos\left(\frac{\pi \omega_2}{3}\right)$$

Putting it all together, a finite closed form expression for v_n can be derived.

$$v_n(x_1, x_2) = -\frac{2^n}{12} \sum_{\omega_1 = -3}^{4} \sum_{\omega_2 = -2}^{3} W(\omega_1, \omega_2), \qquad \text{where}$$

$$W(\omega_1, \omega_2) =$$
$$\cos\left(\pi \left[\frac{x_1 \omega_1}{4} + \frac{x_2 \omega_2}{3}\right]\right) \left[\cos\left(\frac{\pi \omega_1}{4}\right) + \cos\left(\frac{\pi \omega_2}{3}\right)\right]^n \sin\left(\frac{\pi \omega_1}{2}\right) \sin\left(\frac{\pi \omega_2}{3}\right)$$

In the previous example, only moves parallel to a coordinate axis were allowed. These kinds of lattice paths are already well understood in the lattice path literature. On the other hand, the Fourier method is especially useful when diagonal moves are permitted as well.

Example 3.11 Let $\mathbf{h} \in \mathbb{N}^3$. Suppose we want to count paths in the $\mathbf{h}$-corridor allowing only the moves

$$\mathcal{M} = \{(0, \pm 1, 0), (\pm 1, \pm 1, 0), (\pm 1, \pm 1, \pm 1)\},$$

or equivalently to count walks in the graph $G_{\mathbf{h}, \mathcal{M}}$ whose vertices are the points of $C^{\mathbf{h}}$ and whose edges are determined by the move set $\mathcal{M}$. See Fig. 3.9 for the case when $\mathbf{h} = (3, 2, 2)$. Fix an initial point for the walks, $\mathbf{a} = (a_1, a_2, a_3) \in C^{\mathbf{h}}$, and let $\mathbf{d} = (d_1, d_2, d_3) = \mathbf{h} + 1$.

It is straightforward to verify that $T_{\mathcal{M}} = T_2 + T_1 T_2 + T_1 T_2 T_3$.

$$\widehat{T}(\omega) = 2\cos\left(\frac{\pi \omega_2}{d_2}\right) + 4\cos\left(\frac{\pi \omega_1}{d_1}\right)\cos\left(\frac{\pi \omega_2}{d_2}\right) + 8\cos\left(\frac{\pi \omega_1}{d_1}\right)\cos\left(\frac{\pi \omega_2}{d_2}\right)\cos\left(\frac{\pi \omega_3}{d_3}\right)$$
$$(3.26)$$

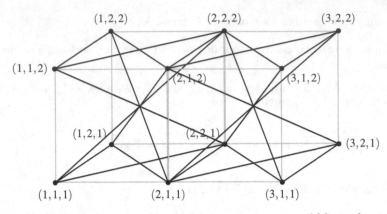

Fig. 3.9 The graph $G_{h,\mathcal{M}}$ corresponding to the fundamental region $\mathcal{C}^{(3,2,2)}$ in $\mathbb{N}^3$, with edges corresponding to move set $\mathcal{M}$ as defined in Example 3.11

Therefore, the state function is:

$$v_n(\mathbf{x}) = \frac{-1}{d_1 d_2 d_3} \sum_{\omega=-\mathbf{d}+1}^{\mathbf{d}} \sin\left(\pi \mathbf{x} \cdot (\omega/\mathbf{d})\right) \left[\widehat{T}(\omega)\right]^n \prod_{j=1}^{3} \sin\left(\frac{\pi \omega_j a_j}{d_j}\right),$$

where $\widehat{T}$ is as defined by Eq. (3.26).

3.4 Walks Within Corridors with Boundary Identifications

Recall from Sect. 2.6, a slight modification of our initial states allowed us to compute the number of walks on a cycle graph, from an initial vertex to any given vertex. Instead of defining v_0 as an *admissible* state, it sufficed to simply define a periodic version of $v_0 : \mathbb{Z} \to \mathbb{Z}$. A cycle graph is nothing more than a path graph whose endpoints are identified; this idea can be generalized.

Let $r \in \mathbb{N}$ and $\mathbf{t} = (t_1, t_2, \ldots, t_r) \in \mathbb{N}^r$. Consider the r-dimensional box.

$$B_{\mathbf{t}} = \prod_{j=1}^{r} \{0, 1, 2, \ldots, t_j\} = \{(x_1, \ldots, x_r) \in \mathbb{Z}^r \mid 0 \le x_j \le t_j, \forall j\} \qquad (3.27)$$

Suppose that a pair of opposite hyperplanes of $B_{\mathbf{t}}$ perpendicular to the x_j-axis are identified. We will call such a set a **corridor with identifications**. State functions v will be required to reflect this identification.

$$v(x_1, \ldots, t_j, \ldots, x_r) = v(x_1, \ldots, 0, \ldots, x_r) \qquad (3.28)$$

It will be fairly straightforward to extend our results from Sect. 2.6 to count walks in graphs whose vertex set is a corridor with identifications. The only modification will be in how the initial state v_0 is defined. Basically, we just have to define v_0 to be periodic of period t_j in the j component. This will ensure that $v = T^n[v_0]$ satisfies Eq. (3.28) for every n, as long as T is a balanced transition operator.

Let us consider the case in which every pair of opposite sides is identified (though our methods can be generalized to any number of identified pairs). That is, we want to count walks in an r-dimensional *torus graph* For simplicity, assume all walks begin at the origin. Then we may define v_0 as the **t**-periodization of δ_0. Just as in Sect. 2.6, the Fourier transform of v_0 is incredibly simple: $V_0(\omega) = 1$.

Theorem 3.3 *Let $r \in \mathbb{N}$ and $\mathbf{t} = (t_1, \ldots, t_r) \in \mathbb{N}^r$. Let $\mathbb{T}$ be the r-dimensional torus graph defined on the points of the box $B_\mathbf{t}$ from Eq. (3.27), with identifications*

$$(x_1, \ldots, t_j, \ldots, x_r) = (x_1, \ldots, 0, \ldots, x_r),$$

*for each j. Given a balanced set of moves $\mathcal{M}$, the nth state for the **toroidal corridor** is obtained by*

$$v_n = \mathcal{F}^{-1}\left[\widehat{T}^n\right],$$

where $\widehat{T}$ is defined as follows.

$$\widehat{T}(\omega) = \sum_{\mathbf{m} \in \mathcal{M}^+} \prod_{j=1}^{r} \left(2\cos\left(\frac{2\pi\omega_j}{t_j}\right)\right)^{m_j}$$

(Note, the Fourier transform and its inverse are defined by (3.10) and (3.11) with periodicity $\mathbf{N} = \mathbf{t}$.)

Example 3.12 Let's derive the state function v_n that counts walks from the origin to a point (x_1, x_2) in the toroidal corridor of dimensions $(t_1, t_2) = (6, 5)$, with move set $\{(\pm 1, 0), (0, \pm 1)\}$ (See Fig. 3.10).

Here, transition operator is $T = T_1 + T_2 = R_1 + R_1^{-1} + R_2 + R_2^{-1}$. Therefore, the function $\widehat{T}$ is given by:

$$\widehat{T}(\omega) = 2\cos\left(\frac{2\pi\omega_1}{t_1}\right) + 2\cos\left(\frac{2\pi\omega_2}{t_2}\right) = 2\left[\cos\left(\frac{\pi\omega_1}{3}\right) + \cos\left(\frac{2\pi\omega_2}{5}\right)\right]$$

Thus, we may construct the state function.

$$v_n(x_1, x_2) = \frac{1}{m_1 m_2} \sum_{\omega_1=0}^{m_1-1} \sum_{\omega_2=0}^{m_2-1} e^{2\pi i\left(\frac{\omega_1 x_1}{t_1} + \frac{\omega_2 x_2}{t_1}\right)} \left(\widehat{T}(\omega)\right)^n$$

$$= \frac{2^n}{30} \sum_{\omega_1=0}^{5} \sum_{\omega_2=0}^{4} e^{2\pi i\left(\frac{\omega_1 x_1}{6} + \frac{\omega_2 x_2}{5}\right)} \left[\cos\left(\frac{\pi\omega_1}{3}\right) + \cos\left(\frac{2\pi\omega_2}{5}\right)\right]^n$$

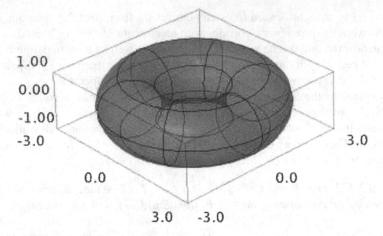

Fig. 3.10 A lattice embedded onto a torus

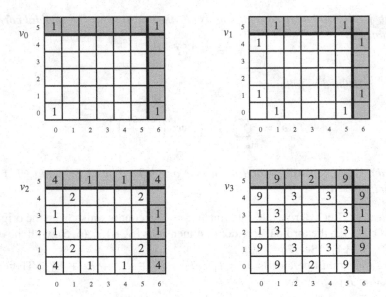

Fig. 3.11 Example corridor state progression $v_0 \rightarrow v_1 \rightarrow v_2 \rightarrow v_3$ in a toroidal corridor with periodicity (6, 5). and transition operator $T = T_1 + T_2$

Finally, because we know that the corridor path counts must be real, we may write:

$$v_n(x_1, x_2) = \frac{2^n}{30} \sum_{\omega_1=0}^{5} \sum_{\omega_2=0}^{4} \cos\left(2\pi \left[\frac{\omega_1 x_1}{6} + \frac{\omega_2 x_2}{5}\right]\right) \left[\cos\left(\frac{\pi \omega_1}{3}\right) + \cos\left(\frac{2\pi \omega_2}{5}\right)\right]^n$$

The first few corridor states for this example are shown in Fig. 3.11.

Exercises

3.1 How many nonidentity moves are possible in $\mathbb{Z}^r$ in general? How many balanced sets of moves are there?

3.2 Fix $r \in \mathbb{N}$. Prove that the shift operators $\{R_1, R_2, \ldots, R_r\}$ (defined on functions $v : \mathbb{Z}^r \to \mathbb{C}$) all commute with one another. That is, show that $R_j R_k = R_k R_j$.

3.3 Prove the composition rule, $H^\eta H^\gamma = H^{\eta\gamma}$. *Hint:* compare $H^\eta [H^\gamma [v]] (\mathbf{x})$ and $H^{\eta\gamma} [v](\mathbf{x})$.

3.4 Mimic the proof of Lemma 3.1 to derive commutation formulae for $H^{\rho_k} L_j$ and $H^{\sigma_k} L_j$. Then use the commutation formulae to prove Lemma 3.2.

3.5 Let $\mathcal{F}$ be the multidimensional DFT as defined by (3.10), and let $\mathcal{F}^{-1}$ denote the inverse DFT as defined by (3.11). Establish the identities $\mathcal{F}\mathcal{F}^{-1} = I$ and $\mathcal{F}^{-1}\mathcal{F} = I$.

3.6 Establish Eq. (3.13).

3.7 Write $\Delta_{(1,2,3)}$ as sums and differences of delta functions (see Example 3.6). Then show that $\Delta_{(1,2,3)}(x_1, x_2, x_3)$ factors as $\Delta_1(x_1)\Delta_2(x_2)\Delta_3(x_3)$.

3.8 Use induction to establish Eq. (3.20).

3.9 Suppose $v(\mathbf{x}) = v_1(x_1)v_2(x_2) \cdots v_r(x_r)$. Let $V_k = \mathcal{F}[v_k]$ for each $1 \le k \le r$.

(a) Show:
$$V(\omega) = \mathcal{F}[v](\omega) = V_1(\omega_1)V_2(\omega_2) \cdots V_r(\omega_r)$$

(b) Show that the above works in reverse as well. That is, if $W(\omega)$ separates as $W_1(\omega_1) \cdots W_r(\omega_r)$, then
$$\mathcal{F}^{-1}[W](\mathbf{x}) = \mathcal{F}^{-1}[W_1](x_1) \cdots \mathcal{F}^{-1}[W_r](x_r)$$

(c) Apply (a) and (b) to compute the DFT of $v(\mathbf{x}) = \Delta_{\mathbf{a}}(\mathbf{x})$, in light of the separation formula (3.20).

3.10 Suppose $U_1, U_2, \ldots, U_r$ are operators and $\widehat{U}_1, \widehat{U}_2, \ldots, \widehat{U}_r$ are functions such that for an arbitrary function v we have $\mathcal{F}[U_j[v]] = \widehat{U}_j \mathcal{F}[v]$. Prove, for any sequence of indices $(k_1, k_2, \ldots, k_m)$ where $1 \le k_j \le r$ for each j, the following holds.
$$\mathcal{F}[U_{k_1} U_{k_2} \cdots U_{k_m}[v]] = \widehat{U}_{k_1} \widehat{U}_{k_2} \cdots \widehat{U}_{k_m} \mathcal{F}[v]$$

3.11 Use (3.21) to find $\widehat{T}(\omega)$ in each case.

(a) $r = 2, \mathbf{d} = (7, 3), \mathcal{M} = \{(0, \pm 1), (\pm 1, \pm 1)\}$
(b) $r = 3, \mathbf{d} = (4, 6, 8), \mathcal{M} = \{(0, 0, 0), (0, 0, \pm 1), (\pm 1, \pm 1, \pm 1)\}$

(c) $r = 4$, $\mathbf{d} = (d_1, d_2, d_3, d_4)$, $\mathcal{M} =$ all moves parallel to an axis
(d) $r \in \mathbb{N}$ arbitrary, $\mathbf{d} = (d_1, \ldots, d_r)$, $\mathcal{M} =$ all moves parallel to an axis

3.12 Finish the proof of Theorem 3.2 by working out the case when r is odd.

3.13 Consider the corridor region $\mathcal{C}^{(6,2)}$ and move set $\mathcal{M} = \{(0, \pm 1), (\pm 1, \pm 1)\}$, as in Exercise 3.11(a).

(a) Produce the corridor state progression for $n = 0, \ldots, 5$ with initial point $\mathbf{a} = (3, 1)$ by hand (i.e. without using Theorem 3.2).
(b) Now use Theorem 3.2 to set up the state function $v_n(\mathbf{x})$ and verify your results from part (a).

3.14 Use Theorem 3.2 and your derivations of $\widehat{T}(\omega)$ from Exercise 3.11 to construct a formula for $v_n(\mathbf{x})$ with an arbitrary initial point $\mathbf{a}$ in the following situations.

(a) $r = 2$, $\mathbf{d} = (7, 3)$, $\mathcal{M} = \{(0, \pm 1), (\pm 1, \pm 1)\}$
(b) $r = 3$, $\mathbf{d} = (4, 6, 8)$, $\mathcal{M} = \{(0, 0, 0), (0, 0, \pm 1), (\pm 1, \pm 1, \pm 1)\}$

Research Questions

3.15 Using mathematical software such as *sage*[1] or another mathematical programming language, create a program that accepts an arbitrary move set $\mathcal{M}$ (checking first whether $\mathcal{M}$ is balanced), corridor dimensions $\mathbf{h}$, and initial point $\mathbf{a} \in \mathcal{C}^{\mathbf{h}}$, and constructs the vertex state function $v_n(\mathbf{x})$. Use your program to produce tables of vertex numbers for various choices of the above parameters. See if there are any interesting patterns in your data.

3.16 Theorem 3.2 requires a balanced set of moves $\mathcal{M}$. However an unbalanced $\mathcal{M}$ can often be analyzed in a lower dimension. For example, the transition operator $T = R_1 R_2 + L_1 L_2$ from Example 3.4 can be understood as the one-dimensional transition operator, $T = R + L$. Study other cases of unbalanced transitions and try to generalize Theorem 3.2.

3.17 Mimic the arguments of Sect. 3.4 to count walks in multidimensional boxes whose opposite faces are identified in the *opposite* way. For example, start by counting walks on a Möbius strip of length d and height h, in which the left and right edges are identified with a *twist*. Position the Möbius strip in the plane with $0 \le x_1 \le d - 1$ and $1 \le x_2 \le h$, as illustrated in Fig. 3.12. Suppose that the initial point is $(0, 1)$. Then the initial state v_0 should be $(2d, 2(h + 1))$-periodic, with the following nonzero values:

$$v_0(0, 1) = v_0(d, h) = 1, \quad \text{and} \quad v_0(0, -1) = v_0(d, -h) = -1$$

---

[1]http://www.sagemath.org/.

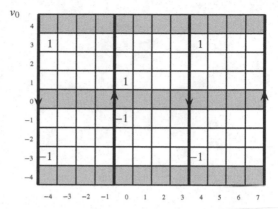

Fig. 3.12 Initial state for a "Möbius strip corridor" of length $d = 4$ and height $h = 3$. Here, walks begin at the point $(0, 1)$. The direction of the arrowheads on the vertical lines suggest the topological identification of the edges of the Möbius strip

3.18 There are various kinds of corridor that can be analyzed using the DFT: bounded, unbounded, and those with certain kinds of identifications on the boundaries. In Chap. 3 we were able to extend our results on bounded corridors to arbitrary dimensions, and Sect. 3.4 generalizes the toroidal case and lays the groundwork for other kinds of pair-wise identifications along the boundary of the corridor. Explore the possibility of developing general formulas that count paths in higher dimensional corridors that exhibit a mixture of these behaviors. In other words, build up a general theory encompassing corridors that are bounded in some directions, unbounded in others, and have pair-wise identification in still others.

Chapter 4
Corridor State Space

So far, in Chaps. 2 and 3, we have introduced and used what are essentially *algebraic* properties of the discrete Fourier transform. In this chapter, we will focus instead on the *geometry* of **corridor state space** (a vector space in which the corridor state vectors v_n can be studied). Geometric properties of the Fourier transform have been widely studied and can be found throughout the general literature.[1]

We begin with results that relate inner products of vertex state vectors. We then explore eigenvector/eigenvalue decomposition of state vectors, which naturally leads the discussion toward the asymptotics of corridor sequences. While the majority of this chapter is devoted to developing the techniques for a certain class of one-dimensional corridors (the so-called *centered* corridors), we also indicate where generalizations may be possible. In particular, we generalize to the multidimensional case in Sect. 4.4 (though still as *centered* corridors). The reader is encouraged to take these ideas even further in the exercises and research questions.

4.1 Vertex States as Vectors

Fix $N \in \mathbb{N}$; and consider the vector space $\mathbb{C}^N$, or the isomorphic subspace of N-periodic vectors of $\mathbb{C}^{\mathbb{Z}}$. Without further ado let us use the notation $\mathbb{C}^N$ for both. So far all that we have used are "algebraic" properties of the Fourier transform (recall Sect. 2.2).

---

[1] A search into the literature for "Fourier transform/methods" produces an overwhelming wealth of information and techniques. Much attention is given to fast computation (central to the development of our digital signal processing world) or abstractions such as generalizations to infinite-dimensional vector spaces. We do not discourage an interested reader to explore Fourier methods further, but advanced concepts are not required in this text.

© Springer Nature Switzerland AG 2019
S. Ault and C. Kicey, *Counting Lattice Paths Using Fourier Methods*, Applied and Numerical Harmonic Analysis,
https://doi.org/10.1007/978-3-030-26696-7_4

- $\mathcal{F} : \mathbb{C}^N \to \mathbb{C}^N$ is a vector space isomorphism, and
- $\mathcal{F}[Rv](\omega) = e^{-\frac{2\pi i \omega}{N}} \mathcal{F}[v](\omega)$, where R is the right-shift operator (our shift, or product, or "convolution" property).

On the other hand, our vector space $\mathbb{C}^N$ also admits *geometric* structure. Ideas such as distance, length, and angle, may be defined via the inner product of complex-valued vectors over the complex field.

$$\langle v, w \rangle = \sum_{x=1}^{N} v(x)\overline{w(x)} \tag{4.1}$$

(See Formula (A.10), and Appendix A.3 in general for a brief overview of complex vector spaces, inner products, and norms.) Note that the complex conjugate in Formula (4.1) is superfluous if $w \in \mathbb{R}^N$. Moreover, the index may be taken over any convenient period $x = x_0, x_0 + 1, \ldots, x_0 + N - 1$. Roughly speaking, the inner product measures the *angle* or amount of *linear independence* between vectors. In particular, for $v, w \in \mathbb{C}^N$, we say that v is orthogonal to w (notated $v \perp w$) if and only if $\langle v, w \rangle = 0$. Note that the inner product formula defines a norm on $\mathbb{C}^N$ via

$$\|v\| = \sqrt{\langle v, v \rangle} = \left(\sum_{x=1}^{N} |v(x)|^2 \right)^{1/2}$$

We require a general version of **Parseval's Theorem** [55].

Theorem 4.1 (Parseval) *Let $v, w \in \mathbb{C}^N$.*

$$\langle v, w \rangle = \frac{1}{N} \langle \mathcal{F}[v], \mathcal{F}[w] \rangle$$

Parseval's Theorem can be used to derive the following "Inner Product Theorem" for corridors. In what follows, we define the inner product of two state vectors v_n and v_m for the (h)-corridor as follows.

$$\langle v_n, v_m \rangle = \sum_{x=1}^{h} v_n(x)v_m(x)$$

In other words, regardless how states are defined on points outside of the fundamental region $\mathcal{C}^{(h)}$, we only care about their values within the corridor for their inner product. Note, because of anti-symmetry of the vertex states, we have $v_n(-x)v_m(-x) = (-v_n(x))(-v_m(x)) = v_n(x)v_m(x)$. Therefore by Parseval's Theorem, we have (keeping in mind that v_m is real, and setting $d = h + 1$ as usual):

$$\langle v_n, v_m \rangle = \frac{1}{2} \sum_{x=-d}^{d-1} v_n(x)v_m(x) = \frac{1}{4d} \sum_{\omega=-d}^{d-1} V_n(\omega)\overline{V_m(\omega)} = \frac{1}{4d} \langle V_n, V_m \rangle \tag{4.2}$$

Theorem 4.2 *Let $h \in \mathbb{N}$ and consider the (h)-corridor $\mathbb{N}_0 \times C^{(h)}$ with initial point $(0, a)$, with either two-way or three-way move sets. For any $n, m \in \mathbb{N}_0$, and $0 \leq j \leq n$, we have:*

$$\langle v_n, v_m \rangle = \langle v_{n-j}, v_{m+j} \rangle \qquad (4.3)$$

Furthermore, we have:

$$\langle v_n, v_m \rangle = v_{m+n}(a) \qquad (4.4)$$

Proof Let $d = h + 1$. First recall from Eqs. (2.20) and (2.28) that $V_n(\omega) = \left(\widehat{T}(\omega)\right)^n V_0(\omega)$, where the transition function on the Fourier side is either $\widehat{T}(\omega) = 2\cos\left(\frac{\pi\omega}{d}\right)$ if paths are two-way (see Eq. (2.17)), or $\widehat{T}(\omega) = 1 + 2\cos\left(\frac{\pi\omega}{d}\right)$ if paths are three-way (see Eq. (2.27)). In either case, $\widehat{T}$ is purely real. Thus we can find a relationship between inner products involving the various vectors $V_n \in \mathbb{C}^{2d}$. For brevity of notation, the input ω is suppressed in functions.

$$\langle V_n, V_m \rangle = \sum_{\omega=-d}^{d-1} \widehat{T}^n V_0 \overline{\widehat{T}^m V_0}$$

$$= \sum_{\omega=-d}^{d-1} \widehat{T}^n V_0 \widehat{T}^m \overline{V_0}$$

$$= \sum_{\omega=-d}^{d-1} \widehat{T}^{n-j} V_0 \widehat{T}^{m+j} \overline{V_0}$$

$$= \sum_{\omega=-d}^{d-1} \widehat{T}^{n-j} V_0 \overline{\widehat{T}^{m+j} V_0}$$

$$= \langle V_{n-j}, V_{m+j} \rangle$$

Next, we use (4.2) to connect this result back to the state vectors v_n.

$$\langle v_n, v_m \rangle = \frac{1}{4d} \langle V_n, V_m \rangle = \frac{1}{4d} \langle V_{n-j}, V_{m+j} \rangle = \langle v_{n-j}, v_{m+j} \rangle$$

This proves (4.3). To prove (4.4), it suffices to observe that $v_0(x) = \delta_a(x)$ in the corridor. Then with $j = n$, we obtain:

$$\langle v_n, v_m \rangle = \langle v_0, v_{m+n} \rangle = \sum_{x=1}^{d-1} \delta_a(x) v_{n+m}(x) = v_{n+m}(a)$$

$\square$

Example 4.1 Consider the corridor shown in Fig. 4.1. Theorem 4.2 holds throughout. For example,

$x = 3$			1	3	8	20	49	119	288	696	1681
$x = 2$		1	2	5	12	29	70	169	408	985	2378
$x = 1$	1	1	2	4	9	21	50	120	289	697	1682
$n =$	0	1	2	3	4	5	6	7	8	9	10

Fig. 4.1 Corridor state progression in $\mathbb{N}_0 \times \mathcal{C}^{(3)}$ with initial point $a = 1$ and three-way moves. Row 1 appears to coincide with OEIS sequence A171842

$$\langle v_4, v_6 \rangle = (9)(50) + (12)(70) + (8)(49) = 1682$$
$$\langle v_3, v_7 \rangle = (4)(120) + (5)(169) + (3)(119) = 1682$$
$$\langle v_2, v_8 \rangle = (2)(289) + (2)(408) + (1)(288) = 1682$$
$$\langle v_1, v_9 \rangle = (1)(697) + (1)(985) + (0)(1681) = 1682$$

All of these agree with $v_{10}(1) = 1682$.

Remark 4.1 Theorem 4.2 can easily be extended to higher dimensional corridors. All that is required is for the transition function $\widehat{T}(\omega)$ to be purely real, which is guaranteed to be the case when the move set under consideration is balanced (see Lemma 3.6 and Eq. (3.21)).

Now suppose that $v \in \mathbb{C}^N$ and $V = \mathcal{F}[v]$ can be decomposed as a linear combination of vectors $\{V^{(1)}, V^{(2)}, \ldots, V^{(r)}\}$ that are pair-wise orthogonal. In other words, there are complex numbers λ_k such that:

$$V = \lambda_1 V^{(1)} + \lambda_2 V^{(2)} + \cdots + \lambda_r V^{(r)} \tag{4.5}$$

Then the Pythagorean Theorem (Theorem A.5) implies:

$$||V||^2 = ||\lambda_1 V^{(1)}||^2 + ||\lambda_2 V^{(2)}||^2 + \cdots + ||\lambda_r V^{(r)}||^2$$
$$= |\lambda_1|^2 ||V^{(1)}||^2 + |\lambda_2|^2 ||V^{(2)}||^2 + \cdots + |\lambda_r|^2 ||V^{(r)}||^2$$

We will use Parseval's Theorem together with the Pythagorean Theorem to derive a useful identity. Recalling Eq. (4.2), consider the squared Euclidean norm of a state vector v_n. Supposing that $V_n = \mathcal{F}[v_n]$ can be decomposed as in (4.5), and further supposing that for all k, we have $||V^{(k)}|| = c$ for some constant c, then:

$$\|v_n\|^2 = \langle v_n, v_n \rangle = \frac{1}{4d} \langle V_n, V_n \rangle = \frac{1}{4d} \|V_n\|^2$$

$$= \frac{1}{4d} \left(|\lambda_1|^2 \|V_1\|^2 + |\lambda_2|^2 \|V_2\|^2 + \cdots + |\lambda_r|^2 \|V_r\|^2 \right)$$

$$= \frac{c^2}{4d} \sum_{k=1}^{r} |\lambda_k|^2 \tag{4.6}$$

The derivations above will help us to develop interesting results about corridor paths in the following sections. In particular, we will find that decompositions like (4.5) exist in which λ_k are the eigenvalues of the corridor state transition functions.

4.2 Using Eigenvectors to Count Corridor Paths

In this section we use some concepts from linear algebra together with the discrete Fourier transform to reinterpret our formulas for certain vertex numbers and corridor numbers with the ultimate goal of deriving approximation formulas (which we will do in Sect. 4.3). Specifically, we use *eigenvector decomposition*; for details, see standard textbooks such as [39, 48]. Recall, if T is a linear operator on a vector space V, and if $v \in V$ is a nonzero vector satisfying $T[v] = \lambda v$ for some scalar λ, then we say v is in **eigenvector** of T with associated **eigenvalue** λ. Then if a given vector w can be decomposed into a linear combination of eigenvectors, v_i with associated eigenvalues λ_i, say $w = c_1 v_1 + \cdots + c_r v_r$, then $T[w]$ becomes easier to compute:

$$T[w] = T \left[\sum_{k=1}^{r} c_k v_k \right] = \sum_{k=1}^{r} c_k T[v_k] = \sum_{k=1}^{r} c_k \lambda_k v_k$$

For our purposes, the vector space V will consist of vertex state functions $v = v_n(x)$ for a given corridor, and we allow complex scalars. Suppose now that v is an eigenvector for a transition function T, with associated eigenvalue $\lambda \in \mathbb{C}$. By the linearity of the DFT, then

$$\widehat{T}(\omega) V(\omega) = \mathcal{F}[T[v]](\omega) = \mathcal{F}[\lambda v](\omega) = \lambda \mathcal{F}[v](\omega) = \lambda V(\omega),$$

so that λ is also an eigenvalue of the operator that is defined as pointwise multiplication by $\widehat{T}(\omega)$, with associated eigenvector being simply $V(\omega)$, the DFT of $v(x)$. Of course this works the other way around too; if λ is an eigenvalue for $\widehat{T}$ with eigenvector V, then λ is also an eigenvalue for the operator T with eigenvector $v = \mathcal{F}^{-1}[V]$.

In fact, eigenvectors for $\widehat{T}$ are trivial to find. Suppose that our vectors are $2d$-periodic for some $d \in \mathbb{N}$. Assume that $\widehat{T}(\omega) = 2\cos(\frac{\pi \omega}{d})$ (the two-way transition function on the Fourier side, as in Sect. 2.3 and Lemma 3.6). Recall that $\widehat{T}$ is $2d$-

periodic and even. For $j \in \{-d, -d + 1, \ldots, d - 2, d - 1\}$, consider the function (vector) $\widetilde{\delta}_j(\omega)$ as defined by Eq. (2.12); that is, $\widetilde{\delta}_j$ is the $2d$-periodization of the delta function δ_j. Pointwise multiplication by $\widehat{T}$ yields the following:

$$\widehat{T}(\omega)\widetilde{\delta}_j(\omega) = \widehat{T}(j)\widetilde{\delta}_j(\omega) = 2\cos\left(\frac{j\pi}{d}\right)\widetilde{\delta}_j(\omega)$$

Thus, $\lambda_j = 2\cos(\frac{j\pi}{d})$ is an eigenvalue for $\widehat{T}$ corresponding to the eigenvector $\widetilde{\delta}_j$. In fact, due to even symmetry, we see that both $\widetilde{\delta}_j$ and $\widetilde{\delta}_{-j}$ are in the eigenspace corresponding to λ_j. These observations will help us to analyze vertex numbers by decomposing the state vectors $v_n = \mathcal{F}^{-1}[V_n]$ into sums of eigenvectors.

While in principle any corridor counting problem can be analyzed using the eigenvector/eigenvalue approach, the computations could be quite tedious. In fact, there is little to no advantage to be had in the general case. On the other hand, there are certain situations in which inherent symmetries make eigenvector analysis the most efficient technique. In what follows, we will use eigenvectors and eigenvalues to study the **centered** corridor problem.

Let us specialize to the following corridor problem: Fix an integer $m \geq 2$, and count the number of n-length walks in the one-dimensional lattice $\mathbb{Z}$ that stay within $m - 1$ steps of the origin, allowing only unit moves left or right. (The reason that we use "$m - 1$" here is to simply formulas going forward.) In other words, we would like to count the number of walks beginning at the center of a fundamental region of the form $\mathcal{H}_{m-1} = \{x \in \mathbb{Z} \mid |x| \leq m - 1\}$, or to count corridor paths in the associated corridor, $\mathbb{N}_0 \times \mathcal{H}_{m-1}$; see Fig. 4.2. As usual, we denote the number of n-length paths ending at (n, x) and staying within the corridor by the state $v_n(x)$, and c_n denotes corridor numbers (total number of n-length corridor paths). We choose to focus on the one-dimensional case for clarity; however, our results can readily generalize to arbitrary dimensions (see Sect. 4.4).

Remark 4.2 This scenario is related to the idea of random walks,[2] as we may wish to compare the number of walks staying within a fixed distance from the origin with the total number of unrestricted walks.

This corridor scenario is in fact equivalent to counting paths in a standard corridor $\mathbb{N}_0 \times \mathcal{C}^{(2m-1)}$, with all paths beginning at $(0, m)$, using the two-way[3] transition operator, $T = L + R$. At this point, we could use Theorem 2.1 directly, but for our purposes in this chapter, we will find it more useful to extract eigenvector/eigenvalue information. The initial state v_0 satisfies $v_0(x) = \delta_m(x)$ for $x \in \mathcal{C}^{(2m-1)}$. Because of the centering, $V_0(\omega)$ has very simple form.

In what follows it may seem at first that we use overly cumbersome notation; however, this is useful as we prepare to generalize to higher dimensions and general balanced operators going forward. First let us define a *parity* function:

---

[2]See e.g. Feller [24] Chap. 14 for a study of random walks from an analytical point of view.

[3]It is quite easy to expand this discussion to the three-way case, but we leave that task as an exercise.

	n=0	1	2	3	4	5	6	7	8	9	10
x = 4											
x = 3				1		4		14		48	
x = 2			1		4		14		48		164
x = 1		1		3		10		34		116	
x = 0	1		2		6		20		68		232
x = −1		1		3		10		34		116	
x = −2			1		4		14		48		164
x = −3				1		4		14		48	
x = −4											

Fig. 4.2 Corridor $\mathbb{N}_0 \times \mathcal{H}_3$ ($m = 4$). The corridor numbers c_n are found by summing the columns: $(1, 2, 4, 8, 14, 28, 48, 96, 164, 328, 560, \dots)$. Note, this is sequence A068912 in the OEIS

$$p(\omega) = \frac{1}{2}(\omega + 1) \tag{4.7}$$

By (2.19), we have:

$$V_0(\omega) = -2i \sin\left(\frac{\pi \omega m}{2m}\right) = -2i \sin\left(\frac{\pi \omega}{2}\right) = \begin{cases} (-1)^{p(\omega)} 2i & \text{if } \omega \text{ is odd,} \\ 0 & \text{if } \omega \text{ is even} \end{cases} \tag{4.8}$$

Observe that V_0 is nonzero only for odd input, and $V_0(-j) = -V_0(j)$.

Next, the two-way transition operator $T = L + R$ corresponds in the usual way to multiplication by $\widehat{T}(\omega) = 2\cos\left(\frac{\pi \omega}{2m}\right)$ on the Fourier side. As above,

$$\lambda_j = \widehat{T}(j) = 2\cos\left(\frac{j\pi}{2m}\right) \tag{4.9}$$

is the eigenvalue for pointwise multiplication by $\widehat{T}(\omega)$ on any vector in the span of $\{\widetilde{\delta}_j, \widetilde{\delta}_{-j}\}$. In particular, λ_j is an eigenvalue for the eigenvector $\Delta_j = \widetilde{\delta}_j - \widetilde{\delta}_{-j}$ (compare Eq. (2.15)). That is,

$$\widehat{T}(\omega)\Delta_j(\omega) = \lambda_j \Delta_j(\omega) \tag{4.10}$$

Now we will build up to our main result slowly, by examining the values of λ_j and V_0. First note that $\lambda_0 = 2$ and $\lambda_{2m} = -2$, regardless of the value of m. Moreover,

Table 4.1 Values of V_0 and $\widehat{T}$ for the centered corridor (one complete period shown)

j	$-2m+1$	...	-5	-4	-3	-2	-1	0	1	2	3	4	5	...	$2m-1$	$2m$
$\widehat{T}(j)$	λ_{2m-1}	...	λ_5	λ_4	λ_3	λ_2	λ_1	2	λ_1	λ_2	λ_3	λ_4	λ_5	...	λ_{2m-1}	-2
$V_0(j)$	$(-1)^m 2i$	...	$2i$	0	$-2i$	0	$2i$	0	$-2i$	0	$2i$	0	$-2i$	...	$(-1)^m 2i$	0

because $\widehat{T}(\omega)$ is an even function of ω, we have $\lambda_{-j} = \lambda_j$. Lining up the values of $\widehat{T}(j) = \lambda_j$ and V_0, a pattern begins to emerge, as suggested by Table 4.1.

Let us examine the structure of $V_0(\omega)$ more closely. With reference to (4.8), $V_0(\omega)$ breaks down as a linear combination of Δ_j functions:

$$V_0(\omega) = -2i\Delta_1(\omega) + 2i\Delta_3(\omega) + \ldots + (-1)^m 2i\Delta_{2m-1}(\omega) \tag{4.11}$$

Let $\mathcal{I}_m = \{1, 3, 5, \ldots, 2m-1\}$. Define for each $j \in \mathcal{I}_m$, the function (vector),

$$V_0^{(j)}(\omega) = (-1)^{p(j)} 2i\Delta_j(\omega) \tag{4.12}$$

Then (4.11) may be expressed more concisely by the following.

$$V_0 = V_0^{(1)} + V_0^{(3)} + V_0^{(5)} + \cdots + V_0^{(2m-1)} = \sum_{j \in \mathcal{I}_m} V_0^{(j)} \tag{4.13}$$

Recall that the values for the DFT of the next state, i.e. $V_1(\omega)$, are found by pointwise multiplication. Therefore, in order to generate V_1, we have the following derivation based on Eqs. (4.10)–(4.13).

$$\begin{aligned}
V_1(\omega) &= \widehat{T}(\omega) V_0(\omega) \\
&= \widehat{T}(\omega) \sum_{j \in \mathcal{I}_m} V_0^{(j)} \\
&= \sum_{j \in \mathcal{I}_m} (-1)^{p(j)} 2i \widehat{T}(\omega) \Delta_j(\omega) \\
&= \sum_{j \in \mathcal{I}_m} (-1)^{p(j)} 2i \lambda_j \Delta_j(\omega) \\
&= \sum_{j \in \mathcal{I}_m} \lambda_j V_0^{(j)}(\omega)
\end{aligned}$$

Iterating the process, with Theorem 3.2 in mind, we have

$$V_n(\omega) = \left[\widehat{T}(\omega)\right]^n V_0(\omega) = \sum_{j \in \mathcal{I}_m} \lambda_j^n V_0^{(j)}(\omega) \tag{4.14}$$

Finally, to obtain the vertex numbers $v_n(x)$, we use linearity of the inverse transform:

$$v_n(x) = \sum_{j \in \mathcal{I}_m} \lambda_j^n \mathcal{F}^{-1}\left[V_0^{(j)} \right](x) = \sum_{j \in \mathcal{I}_m} \lambda_j^n v_0^{(j)}(x), \tag{4.15}$$

where (for odd j),

$$v_0^{(j)}(x) = \frac{(-1)^{p(j)+1}}{m} \sin\left(\frac{j\pi x}{2m} \right) \tag{4.16}$$

(Note, Exercise 4.6 asks you to derive the expression above.)

The preceding discussion proves the following theorem.

Theorem 4.3 *Let $m \in \mathbb{N}$. The state function v_n for the centered corridor $\mathbb{N}_0 \times C^{(2m-1)}$ with two-way paths beginning at $(0, m)$ has the following decomposition.*

$$v_n = \lambda_1^n v_0^{(1)} + \lambda_3^n v_0^{(3)} + \lambda_5^n v_0^{(5)} + \cdots + \lambda_{2m-1}^n v_0^{(2m-1)}, \tag{4.17}$$

where for each odd j, we have

$$\lambda_j = 2\cos\left(\frac{j\pi}{2m} \right), \quad and \quad v_0^{(j)}(x) = \frac{(-1)^{p(j)+1}}{m} \sin\left(\frac{j\pi x}{2m} \right) \tag{4.18}$$

Corollary 4.1 *Under the same assumptions as in Theorem 4.3, we have:*

$$c_n = \frac{1}{m} \sum_{j \in \mathcal{I}_m} (-1)^{p(j)+1} \lambda_j^n S_j,$$

where

$$S_j = \frac{1 + \cos\left(\frac{j\pi}{2m} \right)}{m \sin\left(\frac{j\pi}{2m} \right)} \tag{4.19}$$

Proof See Exercise 4.7 $\qquad\qquad\qquad\qquad\qquad\qquad\qquad\qquad\qquad\qquad\qquad\qquad \square$

Example 4.2 Figure 4.3 shows the centered corridor with size $m = 3$.

For $m = 3$ the eigenvalues are $\lambda_j = 2\cos\left(\frac{j\pi}{6} \right)$, $j = 1, 3, 5$, or $\lambda_1 = \sqrt{3}, \lambda_3 = 0$ and $\lambda_5 = -\sqrt{3}$. Decompose the initial state as $v_0 = v_0^{(1)} + v_0^{(3)} + v_0^{(5)}$ where

$$v_0^{(1)}(x) = \frac{1}{3} \sin\left(\frac{\pi x}{6} \right)$$

$$v_0^{(3)}(x) = -\frac{1}{3} \sin\left(\frac{\pi x}{2} \right)$$

$$v_0^{(5)}(x) = \frac{1}{3} \sin\left(\frac{5\pi x}{6} \right)$$

Then according to Theorem 4.3, this leads to a formula for the corridor vertex states.

	$n=0$	1	2	3	4	5	6	7	8	9	10
$x=6$											
$x=5$			1		3		9		27		81
$x=4$		1		3		9		27		81	
$x=3$	1		2		6		18		54		162
$x=2$		1		3		9		27		81	
$x=1$			1		3		9		27		81
$x=0$											

Fig. 4.3 Centered corridor with size $m=3$, interpreted as $\mathbb{N}_0 \times C^{(7)}$. The corridor numbers are $(c_n)_{n\geq 0} = (1, 2, 4, 6, 12, 18, 36, 54, 108, 162, 324, \ldots)$, which is sequence A068911 in the OEIS

$$
v_n = \begin{bmatrix} v_n(5) \\ v_n(4) \\ v_n(3) \\ v_n(2) \\ v_n(1) \end{bmatrix} = (\sqrt{3})^n \begin{bmatrix} 1/6 \\ \sqrt{3}/6 \\ 1/3 \\ \sqrt{3}/6 \\ 1/6 \end{bmatrix} + 0^n \begin{bmatrix} 1/3 \\ 0 \\ -1/3 \\ 0 \\ 1/3 \end{bmatrix} + (-\sqrt{3})^n \begin{bmatrix} 1/6 \\ -\sqrt{3}/6 \\ 1/3 \\ -\sqrt{3}/6 \\ 1/6 \end{bmatrix} \qquad (4.20)
$$

Notice that the zero eigenvalue only corrects $v_0(x)$, using the convention that $0^0 = 1$. There is an interesting simplification that occurs by looking at only the odd or even length states. For example, if $n = 2\ell > 0$,

$$
v_{2\ell} = (\sqrt{3})^{2\ell} \begin{bmatrix} 1/6 \\ \sqrt{3}/6 \\ 1/3 \\ \sqrt{3}/6 \\ 1/6 \end{bmatrix} + (-\sqrt{3})^{2\ell} \begin{bmatrix} 1/6 \\ -\sqrt{3}/6 \\ 1/3 \\ -\sqrt{3}/6 \\ 1/6 \end{bmatrix} = 3^{\ell-1} \begin{bmatrix} 1 \\ 0 \\ 2 \\ 0 \\ 1 \end{bmatrix}
$$

(The odd case is similar—see Exercise 4.8.) This reflects a natural reduction in the dimension of the *state space* but also can be explained in terms of Fourier analysis (see Exercise 4.11).

Let us also compute the corridor numbers for this scenario for. By Corollary 4.1,

$$
c_n = \frac{1}{3} \left((\sqrt{3})^n \frac{1 + \cos\left(\frac{\pi}{6}\right)}{\sin\left(\frac{\pi}{6}\right)} - (0)^n \frac{1 + \cos\left(\frac{\pi}{2}\right)}{\sin\left(\frac{\pi}{2}\right)} + (-\sqrt{3})^n \frac{1 + \cos\left(\frac{5\pi}{6}\right)}{\sin\left(\frac{5\pi}{6}\right)} \right) \qquad (4.21)
$$

Let's assume that $n \geq 1$, so that only the nonzero eigenvalues will remain. Simplifying (4.21), we have:

$$c_n = \frac{(\sqrt{3})^n}{3} \left(2 + \sqrt{3} + (-1)^n(2 - \sqrt{3})\right) = \begin{cases} 4(3^{\frac{n}{2}-1}) & \text{if } n \text{ is even,} \\ 2(3^{\frac{n-1}{2}}) & \text{if } n \text{ is odd} \end{cases}$$

4.3 Application: Asymptotic Analysis of Centered Corridors

The eigenvector analysis developed in Sect. 4.2 gives exact values for vertex and corridor numbers. When m is small, it is not too much trouble to compute all of the terms of the eigenvalue decomposition; however the advantage in thinking in terms of eigenvalues and eigenvectors is in producing approximate formulas based on the dominant eigenvalues, which are convenient for studying the cases when m is large.

Corollary 4.2 *Under the same assumptions as in Theorem 4.3, we have the following formulas that estimate vertex numbers and corridor numbers for sufficiently large $n \in \mathbb{N}$, in the sense that the ratio of the approximations to exact values approach 1 as $n \to \infty$.*

$$v_n(x) \approx \lambda_1^n v_0^{(1)}(x) + \lambda_{2m-1}^n v_0^{(2m-1)}(x) \tag{4.22}$$

and

$$c_n \approx \lambda_1^n S_1 + (-1)^{m+1} \lambda_{2m-1}^n S_{2m-1}, \tag{4.23}$$

where λ_j and $v_0^{(j)}$ are as in (4.18), and S_j is as in (4.19).

Proof Consider $\lambda_j = 2\cos\left(\frac{j\pi}{2m}\right)$ as a function of j, where $1 \le j \le 2m - 1$. The value of $|\lambda_j|$ is maximized when $j = 1$ and $2m - 1$ (see Fig. 4.4). Thus the dominant eigenvalues are λ_1 and λ_{2m-1}, from which the approximation formulas (4.22) and (4.23) follow. $\qquad\square$

Formulas (4.22) and (4.23) may be simplified using trigonometric identities and splitting into even and odd cases. Exercise 4.9 asks you to do this.

Example 4.3 Consider the $m = 8$ centered corridor. The exact formula for corridor numbers would involve 8 terms (one for each odd integer in $\{1, 3, 5, \ldots, 15\}$). Using Corollary 4.2 to estimate the corridor numbers, we obtain:

$$c_n \approx \left(2\cos\left(\frac{\pi}{16}\right)\right)^n \cdot \frac{1 + \cos\left(\frac{\pi}{16}\right)}{8\sin\left(\frac{\pi}{16}\right)} + (-1)^9 \left(2\cos\left(\frac{15\pi}{16}\right)\right)^n \cdot \frac{1 + \cos\left(\frac{15\pi}{16}\right)}{8\sin\left(\frac{15\pi}{16}\right)}$$

$$= \frac{\left(2\cos\left(\frac{\pi}{16}\right)\right)^n}{8\sin\left(\frac{\pi}{16}\right)} \left[1 + \cos\left(\frac{\pi}{16}\right) - (-1)^n \left(1 - \cos\left(\frac{\pi}{16}\right)\right)\right]$$

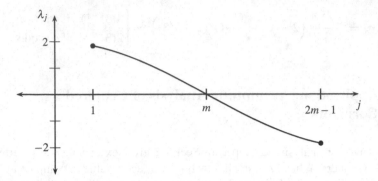

Fig. 4.4 Graph of $\lambda_j = 2\cos(\frac{j\pi}{2m})$ for $1 \le j \le 2m - 1$. The values of $|\lambda_j|$ are maximal at $j = 1$ and $2m - 1$. A continuous graph is shown even though λ_j takes on only discrete values of j in this context

Corollary 4.2 provides one method for estimating vertex and corridor numbers. We may also use *vector norms* to provide bounds on corridor numbers.[4] Recall formula (4.6).

Theorem 4.4 *Under the same assumptions as in Theorem 4.3,*

$$||v_n||^2 = \frac{1}{m} \sum_{j \in \mathcal{I}_m} \lambda_j^{2n} \qquad (4.24)$$

In the above, $\lambda_j = 2\cos(\frac{j\pi}{2m})$ *if counting two-way paths, or* $\lambda_j = 1 + 2\cos(\frac{j\pi}{2m})$ *if counting three-way.*

Proof Recall from (4.14) that V_n may be decomposed as $\sum_{j \in \mathcal{I}_m} \lambda_j^n V_0^{(j)}$. (Throughout this proof, we suppress the input variables in functions.) By (4.12), we have $V_0^{(j)} = (-1)^{p(j)} 2i \Delta_j$, and so $\{V_0^{(1)}, V_0^{(3)}, \ldots, V_0^{(2m-1)}\}$ is a set of mutually orthogonal vectors. Moreover, since Δ_j is nonzero only at two points, taking the values ± 1 at those points, we have $||\Delta_j||^2 = 2$, which implies in turn that $||V_0^{(j)}||^2 = 8$. Use (4.6) with $d = 2m$ and $c^2 = 8$ to derive the result, completing the proof.

$$||v_n||^2 = \frac{8}{8m} \sum_{j \in \mathcal{I}_m} |\lambda_j^n|^2 = \frac{1}{m} \sum_{j \in \mathcal{I}_m} \lambda_j^{2n} \qquad (4.25)$$

(Note that $|\lambda_j^n|^2 = |\lambda_j|^{2n} = \lambda_j^{2n}$ because $\lambda_j \in \mathbb{R}$.)   □

Example 4.4 With $m = 4$, and two-way paths, we have $\lambda_1^2 = \lambda_7^2 = 2 + \sqrt{2}$ and $\lambda_3^2 = \lambda_5^2 = 2 - \sqrt{2}$. So

---

[4]Our discussion below focuses on a relationship between the 1-norm and 2-norm. The latter has its own intrinsic value (e.g., statistical *variance*). The reader is encouraged to think about what other norms may be useful for.

$$\|v_n\|^2 = \frac{1}{4}\left(2\left[2+\sqrt{2}\right]^n + 2\left[2-\sqrt{2}\right]^n\right) = \frac{1}{2}\left(\left[2+\sqrt{2}\right]^n + \left[2-\sqrt{2}\right]^n\right)$$

For large n, the quantity $(2-\sqrt{2})^n$ becomes insignificant since $2-\sqrt{2}<1$. Thus we can get a very good estimate using the dominant eigenvalue.

$$\|v_n\| \approx \sqrt{\frac{1}{2}\left(2+\sqrt{2}\right)^n} = \frac{1}{\sqrt{2}}\left(2+\sqrt{2}\right)^{n/2}$$

4.4   Higher Dimensional Centered Corridors

After developing our tools in the case $r=1$, we are now ready to generalize to $r \in \mathbb{N}$ and centered corridors $\mathbb{N}_0 \times \mathcal{H}_{m_1-1} \times \cdots \times \mathcal{H}_{m_r-1}$, where $\mathbf{m} = (m_1, \ldots, m_r) \in \mathbb{N}^r$ and paths begin at the origin. We will heavily rely on the theory and notation developed in Chap. 3 throughout this section. The theory developed in Sect. 4.2 extends easily to higher dimensions, so we present our results with only minimal discussion or proof.

To begin we consider the equivalent problem of counting walks beginning at $\mathbf{a} = \mathbf{m}$ in the fundamental region $C^{2\mathbf{m}-1}$, where we interpret $2\mathbf{m} - 1 = (2m_1 - 1, 2m_2 - 1, \ldots, 2m_r - 1)$. The corresponding state functions are $4m_k$-periodic in x_k (for all $1 \leq k \leq r$). For this discussion, we assume that the only allowable moves are those parallel to a coordinate axis, and so $T = R_1 + L_1 + \cdots + R_r + L_r = T_1 + \cdots + T_r$ (but in principle, any balanced operator could be considered). Thus $\widehat{T}(\omega) = 2\sum_{k=1}^{r}\cos(\frac{\pi\omega_k}{2m_k})$.

Let $\mathbf{j} = (j_1, j_2, \ldots, j_r) \in \mathbb{Z}^r$. As in Sect. 4.2, we have the key observations that define our eigenvectors and eigenvalues (compare Eqs. (4.9) and (4.10)).

$$\lambda_{\mathbf{j}} = \widehat{T}(\mathbf{j}) = 2\sum_{k=1}^{r}\cos\left(\frac{j_k\pi}{2m_k}\right) \tag{4.26}$$

is the eigenvalue for pointwise multiplication by $\widehat{T}(\omega)$ on the eigenvector $\Delta_{\mathbf{j}}(\omega)$. That is,

$$\widehat{T}(\omega)\Delta_{\mathbf{j}}(\omega) = \lambda_{\mathbf{j}}\Delta_{\mathbf{j}}(\omega) \tag{4.27}$$

The DFT of the initial state is

$$V_0(\omega) = \mathcal{F}[\Delta_{\mathbf{m}}](\omega) = (-2i)^r\prod_{k=1}^{r}\sin\left(\frac{\pi\omega_k}{2}\right) \tag{4.28}$$

Note that (4.28) does not depend on $\mathbf{m}$. Observe also that V_0 is nonzero only when all components of the input are odd. Extend the parity function (4.7) to arbitrary $\omega \in \mathbb{Z}^r$ via:

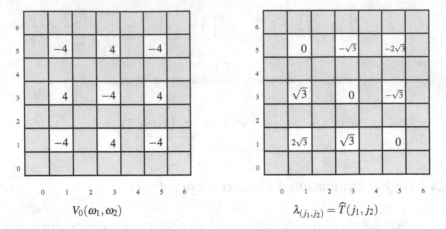

$$V_0(\omega_1, \omega_2) \qquad\qquad\qquad \lambda_{(j_1, j_2)} = \widehat{T}(j_1, j_2)$$

Fig. 4.5 The table on the left shows the nonzero values of $V_0(\omega_1, \omega_2)$ in $\mathcal{I}_{\mathbf{m}}$ for $r = 2$ and $\mathbf{m} = (3, 3)$. These values extend by antisymmetry across each boundary. The table on the right shows the corresponding values of $\lambda_{(j_1, j_2)}$. The values corresponding to position (j_1, j_2) in which either j_k is even are irrelevant. These values extend by even symmetry across each boundary

$$p(\omega) = \frac{1}{2}(\omega_1 + \omega_2 + \cdots + \omega_r + r)$$

Then (4.28) reduces to

$$V_0(\omega) = \begin{cases} (-1)^{p(\omega)}(2i)^r & \text{if all } \omega_1, \omega_2, \ldots, \omega_r \text{ are odd,} \\ 0 & \text{otherwise (i.e., any } \omega_k \text{ is even)} \end{cases} \qquad (4.29)$$

Let $\mathcal{I}_{\mathbf{m}} = \mathcal{I}_{m_1} \times \cdots \times \mathcal{I}_{m_r}$, and let $\mathbf{j} \in \mathcal{I}_{\mathbf{m}}$. We now decompose $V_0(\omega)$ in terms of the eigenfunctions $\Delta_{\mathbf{j}}$.

$$V_0(\omega) = \sum_{\mathbf{j} \in \mathcal{I}_{\mathbf{m}}} V_0^{\mathbf{j}}(\omega),$$

where

$$V_0^{\mathbf{j}}(\omega) = V_0(\mathbf{j})\Delta_{\mathbf{j}}(\omega) = (-1)^{p(\mathbf{j})}(2i)^r \Delta_{\mathbf{j}}(\omega) \qquad (4.30)$$

Hence each $V_0^{\mathbf{j}}$ is also an eigenfunction for the operation of multiplication by $\widehat{T}(\omega)$ with associated eigenvalue $\lambda_{\mathbf{j}}$ as defined by Eq. (4.26). For example, see Fig. 4.5.

We are now ready to generalize Theorem 4.3.

Theorem 4.5 *Let* $r \geq 1$ *and* $\mathbf{m} = (m_1, \ldots, m_r) \in \mathbb{N}^r$. *The number of n-length walks in the fundamental region* $\mathcal{C}^{2\mathbf{m}-1}$ *starting at the center point* $\mathbf{m}$ *and ending at* $\mathbf{x}$, *with moves parallel to the coordinate axes, is given by the state function,*

$$v_n(\mathbf{x}) = \sum_{\mathbf{j} \in \mathcal{I}_{\mathbf{m}}} \lambda_{\mathbf{j}}^n v_0^{\mathbf{j}}(\mathbf{x}),$$

where

$$\lambda_{\mathbf{j}} = 2 \sum_{k=1}^{r} \cos\left(\frac{j_k \pi}{2m_k}\right), \quad and \quad v_0^{\mathbf{j}}(\mathbf{x}) = (-1)^{p(\mathbf{j})+r} \prod_{k=1}^{r} \frac{1}{m_k} \sin\left(\frac{j_k \pi x_k}{2m_k}\right)$$

Proof The multidimensional analog of Eq. (4.14) is as follows:

$$V_n(\omega) = \sum_{\mathbf{j} \in \mathcal{I}_{\mathbf{m}}} \lambda_{\mathbf{j}}^n V_0^{\mathbf{j}}(\omega). \tag{4.31}$$

Using Theorem 3.2, we simply apply the inverse Fourier transform to the function V_n to obtain v_n. All that remains is to find $v_0^{\mathbf{j}}$.

$$v_0^{\mathbf{j}} = \mathcal{F}^{-1}[V_0^{\mathbf{j}}] = (2i)^r (-1)^{p(\mathbf{j})} \mathcal{F}^{-1}[\Delta_{\mathbf{j}}] \tag{4.32}$$

$$v_0^{\mathbf{j}}(\mathbf{x}) = (2i)^r (-1)^{p(\mathbf{j})} \cdot \frac{i^r}{2^r} \prod_{k=1}^{r} \frac{1}{m_k} \sin\left(\frac{j_k \pi x_k}{2m_k}\right) \tag{4.33}$$

The result follows after simplifying. $\qquad\qquad\square$

A routine calculation yields the corridor numbers c_n by summing the $v_n(x)$ from Theorem 4.5 over the fundamental region $C^{2\mathbf{m}-1}$. We state the result in terms of centered corridors. Define for $\mathbf{j} = (j_1, \ldots, j_r)$,

$$S_{\mathbf{j}} = \prod_{k=1}^{r} \frac{1 + \cos\left(\frac{j_k \pi}{2m_k}\right)}{m_k \sin\left(\frac{j_k \pi}{2m_k}\right)} \tag{4.34}$$

Theorem 4.6 *Let $r \geq 1$ and $\mathbf{m} = (m_1, \ldots, m_r) \in \mathbb{N}^r$. The number of n-length walks starting at the origin and staying within $\mathcal{H}_{m_1-1} \times \cdots \times \mathcal{H}_{m_r-1}$, allowing only unit moves in directions parallel to the coordinate axes, is given by*

$$c_n = \sum_{\mathbf{j} \in \mathcal{I}_{\mathbf{m}}} (-1)^{p(\mathbf{j})+r} \lambda_{\mathbf{j}}^n S_{\mathbf{j}} \tag{4.35}$$

Proof The proof is a straightforward generalization of the proof of Corollary 4.1.

Fortunately all sums are finite, so issues of convergence do not arise. Exercise 4.14 asks you to fill in a missing algebraic step. $\qquad\qquad\square$

Remark 4.3 Theorems 4.5 and 4.6 may be generalized even further to encompass arbitrary balanced move sets. All that needs to change are the eigenvalues $\lambda_{\mathbf{j}} = \widehat{T}(\mathbf{j})$, where $\widehat{T}(\omega)$ corresponds to a balanced transition operator T.

Example 4.5 Consider the move set $\mathcal{M} = \{(\pm 1, 0), (\pm 1, \pm 1)\}$. The balanced transition operator is $T = T_1 + T_1 T_2$. Therefore, our eigenvalues are:

	1	2	3	4	5
3	$\sqrt{3}-\sqrt{6}$		0		$-\sqrt{3}+\sqrt{6}$
2					
1	$\sqrt{3}+\sqrt{6}$		0		$-\sqrt{3}-\sqrt{6}$

Fig. 4.6 Important values of $\lambda_{(j_1,j_2)} = 2\cos(\frac{j_1\pi}{6}) + 4\cos(\frac{j_1\pi}{6})\cos(\frac{j_1\pi}{4})$

$$\lambda_{(j_1,j_2)} = 2\cos\left(\frac{j_1\pi}{2m_1}\right) + 4\cos\left(\frac{j_1\pi}{2m_1}\right)\cos\left(\frac{j_1\pi}{2m_2}\right)$$

With $(m_1, m_2) = (3, 2)$, there would be six eigenvalue/eigenvector pairs, but we find that two are zero; see Fig. 4.6.

Observe that $|\lambda_{(1,1)}| = |\lambda_{(5,1)}| = \sqrt{3}+\sqrt{6}$ are the dominant eigenvalues. A multidimensional version of Theorem 4.4 would imply:

$$||v_n||^2 = \frac{1}{3\cdot 2}\left[2(\sqrt{3}+\sqrt{6})^{2n} + 2(0)^{2n} + 2(\sqrt{3}-\sqrt{6})^{2n}\right]$$
$$\approx \frac{1}{3}(\sqrt{3}+\sqrt{6})^{2n} = \frac{1}{3}(9+6\sqrt{2})^n$$

This example may lead one to consider more general situations. Research Question 4.17 asks you to explore balanced move sets in more generality.

What is particularly nice about Theorems 4.5 and 4.6 is that these results provide means by which the vertex numbers and corridor numbers may be approximated by simpler formulae. For large n, only the dominant eigenvalue(s) are required for asymptotic analysis of $v_n(\mathbf{x})$ or c_n, as we did in Example 4.5 above. Presently we give the explicit result for the centered corridor numbers.

Corollary 4.3 *Write* $\mathbf{1} = (1, 1, \ldots, 1) \in \mathbb{N}^r$. *Under the same assumptions as in Theorem 4.6, we have:*

$$c_n \approx \lambda_{\mathbf{1}}^n S_{\mathbf{1}} + (-1)^{r+\sum_{k=1}^r m_k}\lambda_{2\mathbf{m}-1}^n S_{2\mathbf{m}-1}, \tag{4.36}$$

where $\lambda_{\mathbf{j}}$ *as in (4.26), and* $S_{\mathbf{j}}$ *is as in (4.34).*

Proof The dominant eigenvalues occur when $\mathbf{j} = (1, \ldots, 1)$ and $\mathbf{j} = (2m_1 - 1, \ldots, 2m_r - 1)$, at which points we have $|\lambda_{\mathbf{j}}|$ as close to the value $2r$ as possible within the restricted domain $1 \le j_k \le 2m_k - 1$. (See Fig. 4.7 for a two-dimensional example.) □

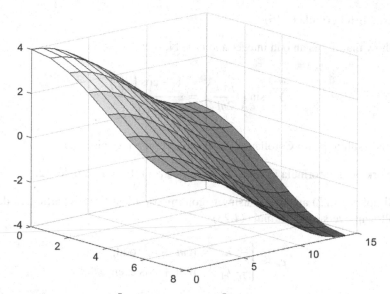

Fig. 4.7 Graph of $\lambda_{(j_1, j_2)} = 2\left[\cos(\frac{j_1\pi}{2m_1}) + \cos(\frac{j_2\pi}{2m_2})\right]$ for $(m_1, m_2) = (4, 7)$. We consider only lattice points (j_1, j_2) such that $1 \leq j_k \leq m_k$. Within this restricted domain, the values of $|\lambda_{(j_1, j_2)}|$ are maximal at $j = (1, 1)$ and $(2m_1 - 1, 2m_2 - 1) = (7, 13)$

Exercises

4.1 Let $v = (1, 1, -1, -1)$ and $w = (i, 1, -i, -1)$, indexed starting at 0 and extended by $N = 4$ periodicity.

(a) Compute the DFTs, $V = \mathcal{F}[v]$ and $W = \mathcal{F}[w]$.
(b) Verify the shift property: $\mathcal{F}[R[v]](\omega) = e^{-\frac{2\pi i \omega}{4}} V(\omega)$.
(c) Verify that: $\langle v, w \rangle = \frac{1}{4}\langle V, W \rangle$.

4.2 Let $N \in \mathbb{N}$ and suppose that $v = (v(x))_{x=-\infty}^{\infty}$ is $2N$-periodic and *antisymmetric* (i.e. $v(-x) = -v(x)$ for all $x \in \mathbb{Z}$).

(a) Show that v must satisfy $v(qN) = 0$ for all $q \in \mathbb{Z}$.
(b) Show that $V(\omega)$ must be purely imaginary.

4.3 Let $N \in \mathbb{N}$. For $k = 0, 1, \ldots, N - 1$, let $e_k(x) = e^{-\frac{2k\pi i x}{N}}$. Show that the vectors $e_0, e_1, \ldots, e_{N-1} \in \mathbb{C}^N$ defined by $e_k = (e_k(x))_{x=0}^{N-1}$ are pairwise orthogonal.
 Then use this result to prove Theorem 4.1.

4.4 By hand, verify that $\langle v_4, v_2 \rangle = v_6(a_1, a_2)$ for the corridor $\mathbb{N}_0 \times \mathcal{C}^{(4,3)}$ using move set $\mathcal{M} = \{(\pm 1, 0), (\pm 1, \pm 1)\}$ and initial point $(a_1, a_2) = (2, 1)$.

4.5 Let $m \in \mathbb{N}$. Show that $\widehat{T}(\omega) = 2\cos\left(\frac{\pi\omega}{2m}\right)$ is antisymmetric about $\omega = m$, i.e., $\widehat{T}(\omega) = -\widehat{T}(2m - \omega)$.

4.6 Establish Formula (4.16).

4.7 Show that if j is an odd integer and $m \in \mathbb{N}$, then

$$\sum_{x=1}^{2m-1} \sin\left(\frac{j\pi x}{2m}\right) = \frac{1 + \cos\left(\frac{j\pi}{2m}\right)}{\sin\left(\frac{j\pi}{2m}\right)}$$

Use this result to prove Corollary 4.1. *Hint:* Compare Exercise A.5.

4.8 Work out the formula for the odd case, $v_{2\ell+1}(x)$, from Example 4.2.

4.9 Simplify (4.22) and (4.23) using trigonometric identities. In particular, derive the following version of formula (4.23):

$$c_n \approx \begin{cases} \frac{2}{D_m} \lambda_1^n & \text{if } m + n \text{ is odd,} \\ \frac{1}{D_m} \lambda_1^{n+1} & \text{if } m + n \text{ is even} \end{cases}$$

Here, $\lambda_1 = 2\cos\left(\frac{\pi}{2m}\right)$, and $D_m = m\sin\left(\frac{\pi}{2m}\right)$.

4.10 Extend Theorem 4.3 to cover the three-way case ($T = L + I + R$).

4.11 Let $v_n(x)$ be the state vectors for the centered corridor problem with $m = 4$. (two-way paths in $\mathbb{N}_0 \times C^{(7)}$ starting at $(0, 4)$).

(a) Decomposing $V_0(\omega) = W^{(1)}(\omega) + W^{(3)}(\omega)$, where

$$W^{(1)}(\omega) = -2i \left[\Delta_1(\omega) - \Delta_7(\omega)\right] \quad \text{and} \quad W^{(3)}(\omega) = 2i \left[\Delta_3(\omega) - \Delta_5(\omega)\right],$$

find the appropriate eigenvalues μ_1, μ_3 such that

$$V_{2n}(\omega) = \mu_1^n W^{(1)}(\omega) + \mu_3^n W^{(3)}(\omega)$$

Hint: We are looking for eigenvalues for the *squared* transition function $\widehat{T}^2$. Begin with a table like Table 4.1 but with $\mu_j = \widehat{T}^2(j)$ (only $\mu_1, \ldots, \mu_{m-1}$ need to appear).

(b) By taking inverse Fourier transforms, write $v_{2n}(x) = \mu_1^n w^{(1)}(x) + \mu_3^n w^{(3)}(x)$, for appropriate functions $w^{(1)}$ and $w^{(3)}$. Create a similar decomposition for the corridor numbers c_{2n} in the form $c_{2n} = \mu_1^n c_1 + \mu_3^n c_3$ for appropriate values c_1 and c_3.

(c) With the help of a computer, compare the approximation $c_{2n} \approx \mu_1^n c_1$ to c_{2n} itself.

(d) Now repeat the above steps, but with three-way paths.

4.12 What changes in Corollary 4.2 when the transition function is changed to $T = L + R + I$ (three-way moves)? *Hint:* Where is/are the dominant eigenvalue(s) of the form $\lambda_j = \widehat{T}(j) = 1 + 2\cos(\frac{j\pi}{2m})$?

4.13 Verify the following equation in order to fill in the missing details for (4.32).

$$\mathcal{F}^{-1}[\Delta_j] = \frac{i^r}{2^r} \prod_{k=1}^{r} \frac{1}{m_k} \sin\left(\frac{j_k \pi x_k}{2m_k}\right)$$

4.14 Prove of Theorem 4.6 in detail, by first establishing the following key result
Fix $\mathbf{a} = (a_1, \ldots, a_r) \in \mathbb{N}^r$, and suppose $f(a, x)$ is defined for all $a, x \in \mathbb{Z}$. Let $\mathbf{h} = (h_1, \ldots, h_r) \in \mathbb{N}^r$. By inductively factoring, prove:

$$\sum_{\mathbf{x} \in C^{\mathbf{h}}} \prod_{k=1}^{r} f(a_k, x_k) = \left[\sum_{x_1=1}^{h_1} f(a_1, x_k)\right]\left[\sum_{x_2=1}^{h_2} f(a_2, x_k)\right] \cdots \left[\sum_{x_r=1}^{h_r} f(a_r, x_k)\right]$$

4.15 Simplify (4.36) (from Corollary 4.3) as much as possible.

4.16 Consider the basic random walk with $Y_0 = 0$. Find the proportion of walks $Y_0, Y_1, \ldots, Y_{2n}$ so that

(a) $Y_\ell \in \{-2, -1, 0, 1, 2\}$ for all $\ell \leq 2n$, versus
(b) the walk ends with $Y_{2n} \in \{-2, -1, 0, 1, 2\}$.

Research Questions

4.17 Extend the eigenvector/eigenvalue methods developed in this chapter to work with general balanced move sets.

4.18 Obtain error bounds for the approximation formulas for c_n given in Formulas (4.23) and (4.36).

4.19 Consider the basic random walk $Y_0, Y_1, \ldots$ with $Y_0 = 0$, moving up or down at each discrete time step with probability 1/2. Also consider a three-way random walk $X_0, X_1, \ldots$ with $X_0 = 0$, moving up or down or not moving at all, at each discrete time step with probability 1/3. Do you think it is more likely that a random walk of the first type or the random walk of the second type would remain in $\mathbb{N}_0 \times \{-(m-1), \ldots, 0, \ldots, m-1\}$? (We found for $m = 3$ and $n = 6$, the first type has probability $9/16 = 0.5625$, while second type has probability $517/3^6 \approx 0.7092$.) Compute or estimate these probabilities for general $m, n \in \mathbb{N}$.

4.20 How would Theorems 4.5 and 4.6 change if the initial point of paths were not $\mathbf{m} \in C^{2\mathbf{m}}$? In other words, what is the eigenvector/eigenvalue analysis for *uncentered* corridors?

Appendix A
Review: Complex Numbers

Our Fourier methods for counting paths in corridors require familiarity with complex numbers and functions. What follows is a concise primer in complex arithmetic and the exponential function, which is standard and can be found in numerous elementary texts including [11, 13, 52]. Further theory and applications concerning Fourier transforms may be found in [21, 51, 55].

A.1 Properties of the Complex Numbers

We denote the set of **complex numbers** by $\mathbb{C} = \{x + yi : x, y \in \mathbb{R}\}$, where $i^2 = -1$, i.e., $i = \sqrt{-1}$. Given $z = x + yi \in \mathbb{C}$ with $x, y \in \mathbb{R}$, we say that z is expressed in **standard** (or **rectangular**) form. Moreover, we say that x is the **real part** and y is the **imaginary part** of z, denoted $x = \operatorname{Re} z$ and $y = \operatorname{Im} z$, respectively. Henceforth when we write $z = x + yi$, we mean that z is in standard form (and so $x, y \in \mathbb{R}$ is presumed).

The standard form $z = x + yi$ suggests that z can be interpreted as a point $(x, y) \in \mathbb{R}^2$. In fact, each point $(x, y) \in \mathbb{R}^2$ determines the associated complex number $z = x + yi$ uniquely. In this context, $\mathbb{R}^2$ will be called the **complex plane**. Moreover, it is often convenient to interpret $z = x + yi$ as a *vector* $(x, y) \in \mathbb{R}^2$, that is, the vector that can be represented by starting at the origin and ending at (x, y). Under this interpretation, $\mathbb{C}$ becomes a **vector space**.[1]

For $z_1 = x_1 + y_1 i$ and $z_2 = x_2 + y_2 i$ and $c \in \mathbb{R}$, we define complex addition by $z_1 + z_2 = (x_1 + x_2) + (y_1 + y_2)i$ and scalar multiplication by $cz_1 = (cx_1) + (cy_1)i$. Thus, complex addition and multiplication by a real number can be exactly interpreted as usual vector addition and scalar multiplication in $\mathbb{R}^2$. Subtraction of complex

---

[1]In fact, $\mathbb{C}$ is simultaneously a two-dimensional *real* vector space and one-dimensional *complex* vector space. The choice of scalars matters and must be fixed beforehand.

© Springer Nature Switzerland AG 2019

S. Ault and C. Kicey, *Counting Lattice Paths Using Fourier Methods*, Applied and Numerical Harmonic Analysis,
https://doi.org/10.1007/978-3-030-26696-7

numbers may be defined by $z_1 - z_2 = z_1 + (-1)z_2$. The **length** or **magnitude** of $z = x + yi$ is defined by $|z| = \sqrt{x^2 + y^2}$. Geometrically, $|z| \geq 0$ is simply the distance from the origin to (x, y) in $\mathbb{R}^2$ or the length of the vector $(x, y) \in \mathbb{R}^2$. As such, the complex plane $\mathbb{C}$ may be regarded as the familiar Euclidean space $\mathbb{R}^2$ with distance or **metric** function $d(z_1, z_2) = |z_1 - z_2| = \sqrt{(x_1 - x_2)^2 + (y_1 - y_2)^2}$.

If Im $z = 0$, then usually we write $z = x + 0i = x$ and so $\mathbb{R} \subseteq \mathbb{C}$, with $\mathbb{R}$ corresponding geometrically to the (horizontal) x-axis of $\mathbb{R}^2$, what we call the **real axis** in the complex plane. If Re $z = 0$, we say that $z = 0 + yi = yi$ is **purely imaginary** and yi corresponds to the point $(0, y)$ on the (vertical) y-axis or **imaginary axis** of the complex plane.

If $z = x + yi \in C$ we define the (complex) **conjugate** $\bar{z}$ of z by $\bar{z} = x - yi$. Geometrically, if we interpret z as a vector, then $\bar{z}$ is the reflection or flip of z over the real axis. For $z_1 = x_1 + y_1 i$ and $z_2 = x_2 + y_2 i$, define complex multiplication as follows.

$$z_1 z_2 = (x_1 x_2 - y_1 y_2) + (x_1 y_2 + x_2 y_1)i \tag{A.1}$$

Note that with this definition, for all $z \in \mathbb{C}$, we have $z\bar{z} = |z|^2$. Also note that this definition is consistent with the real number multiplication (i.e., when $y_1 = y_2 = 0$, Eq. (A.1) reduces to $x_1 x_2$). We will say a little bit about the geometric interpretation of complex multiplication shortly. With the addition and multiplications thus defined, $\mathbb{C}$ is a *commutative field*[2] with additive identity $0 = 0 + 0i$ and multiplicative identity $1 = 1 + 0i$. In particular, every complex number $z \neq 0$ has a **multiplicative inverse**, z^{-1}. Indeed, if $z = x + yi$, then the inverse is given by:

$$z^{-1} = \frac{1}{z} = \frac{1}{z}\frac{\bar{z}}{\bar{z}} = \left(\frac{1}{|z|^2}\right)\bar{z} = \left(\frac{x}{x^2 + y^2}\right) + \left(\frac{-y}{x^2 + y^2}\right)i \tag{A.2}$$

Now we may define complex division. If $z_1, z_2 \in \mathbb{C}$ and $z_2 \neq 0$, then define $z_1/z_2 = z_1(z_2^{-1})$. For $n \in \mathbb{N}_0$, define the powers z^n inductively by $z^{n+1} = z^n z$ and $z^0 = 1$. Also define $z^{-n} = (z^{-1})^n$. For *integer* exponents, we may use induction to develop the laws of exponents. In what follows we will not need to deal with the subtleties, such as branches, encountered with noninteger exponents.

Theorem A.1 *Let $z, w \in \mathbb{C}$ and $m, n \in \mathbb{Z}$.*

(a) $z^m z^n = z^{m+n}$ (g) $z - \bar{z} = 2\,\text{Im}(z)i$

(b) $(z^m)^n = z^{mn}$ (h) $z \in \mathbb{R} \iff \bar{z} = z$

(c) $\overline{z + w} = \bar{z} + \bar{w}$ (i) $|z| \geq 0$

(d) $\overline{zw} = \bar{z}\,\bar{w}$ (j) $|z| = 0 \iff z = 0$

(e) $\bar{\bar{z}} = z$ (k) $|zw| = |z|\,|w|$

(f) $z + \bar{z} = 2\,\text{Re}(z)$ (l) $|z + w| \leq |z| + |w|$

Proof All are straightforward to prove except for *(l)*, which is the famous and useful **triangle inequality** in $\mathbb{R}^2$.   □

---

[2]Thorough knowledge of field theory is not required for the material in this book.

Finally, note that typical proofs of the Binomial Theorem (see Theorem 1.1) and formula for the sum of the geometric series, which only require the algebraic field properties of $\mathbb{R}$, extend easily to complex numbers. In particular, if $z \in \mathbb{C}$ with $z \neq 1$, then for all $n \in \mathbb{N}$, we have:

$$\sum_{k=0}^{n-1} z^k = 1 + z + z^2 + \cdots + z^{n-1} = \frac{1 - z^n}{1 - z} \tag{A.3}$$

Note that when $z = 1$, the left-hand side is simply equal to n.

A.2 The Complex Exponential

Recall the (real) exponential function, $f(x) = e^x$. There are many ways to extend the definition of the exponential to complex numbers. Our purpose is not to rigorously define the *complex exponential*, but instead to rely on a particular formula to define it, called **Euler's Formula**.

Every point (or vector) $(x, y) \in \mathbb{R}^2$ can be expressed in polar coordinates by $(x, y) = (r \cos \theta, r \sin \theta)$ for an appropriate choice of real r and θ. Thus if $z = x + yi \in \mathbb{C}$, then we have $x = r \cos \theta$ and $y = r \sin \theta$. Consider $r = 1$. Then $z = \cos \theta + (\sin \theta)i$ is on the unit circle, and we *define* the **complex exponential** for every angle $\theta \in \mathbb{R}$ by:

$$Euler's\ Formula: \quad e^{i\theta} = \cos \theta + i \sin \theta \tag{A.4}$$

Note that the placement of the "i" occurs typically in front of θ in the exponent rather than behind it as standard form would seem to dictate (the same juxtaposition also shows up with regard to $\sin \theta$). The reason for this slight inconsistency in notation is simply mathematical tradition. Thus, for $\theta \in \mathbb{R}$, $e^{i\theta}$ is just a point on the unit circle, or a unit vector from the origin forming an angle θ with the positive real axis in the complex plane (Fig. A.1).

Then for any $z \in \mathbb{C}$, we have $z = r \cos \theta + ir \sin \theta = r(\cos \theta + i \sin \theta) = re^{i\theta}$. Indeed, $r = \pm |z|$ may be regarded as simply a scaling factor on the unit vector $e^{i\theta}$. When written in the form $z = re^{i\theta}$, we say that z is in **polar form**.

Remark A.1 More generally, e^z may be defined for $z = x + yi \in \mathbb{C}$ (where $x, y \in \mathbb{R}$) by $e^{x+yi} = e^x e^{iy} = e^x(\cos y + i \sin y)$. As a consequence of this definition, e^z is periodic with period $2\pi i$.

Moreover, since cosine is an even function and sine is an odd function, from (A.4), we have $e^{-i\theta} = e^{i(-\theta)} = \cos(-\theta) + i \sin(-\theta)$, or $e^{-i\theta} = \cos \theta - i \sin \theta$, from which **Euler's Identities** for sine and cosine follow.

$$\cos \theta = \frac{1}{2}\left(e^{i\theta} + e^{-i\theta}\right) \quad \text{and} \quad \sin \theta = \frac{1}{2i}\left(e^{i\theta} - e^{-i\theta}\right) \tag{A.5}$$

Fig. A.1 The complex plane. Every number $z = x + yi = r\cos\theta + (r\sin\theta)i = re^{i\theta} \in \mathbb{C}$ is represented by the point (x, y) in rectangular coordinates, or (r, θ) in polar coordinates

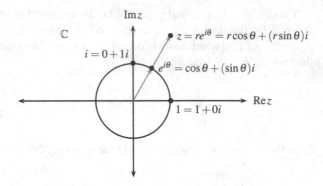

Further important properties of the complex exponential are provided below.

Theorem A.2 *Let* $\theta, \theta_1, \theta_2 \in \mathbb{R}$.

(a) $e^{i\theta_1} e^{i\theta_2} = e^{i(\theta_1 + \theta_2)}$

(b) $e^{i\theta_1} / e^{i\theta_2} = e^{i(\theta_1 - \theta_2)}$

(c) $e^{-i\theta} = \overline{e^{i\theta}}$

(d) $e^{i\theta_1} = e^{i\theta_2} \iff \theta_1 - \theta_2 = 2\pi k$, *for some* $k \in \mathbb{Z}$

Now suppose $z_1, z_2 \in \mathbb{C}$ are expressed in polar coordinates; $z_k = r_k e^{i\theta_k}$ for $k = 1, 2$. Then $z_1 z_2 = r_1 r_2 e^{i(\theta_1 + \theta_2)}$, which gives a geometric interpretation of complex multiplication: To multiply two complex numbers (vectors), simply multiply their lengths and add their polar angles. Observe that if $\theta_1 = \theta_2 = \theta$ in Theorem A.2(a), we have $(e^{i\theta})^2 = e^{i(2\theta)}$. By induction, one can easily prove the following.

Theorem A.3 (De Moivre) *Let* $\theta \in \mathbb{R}$ *and* $n \in \mathbb{Z}$. *Then* $\left(e^{i\theta}\right)^n = e^{in\theta}$.

Alternatively, we may express the previous result together with Euler's Formula.

$$(\cos\theta + i\sin\theta)^n = \cos(n\theta) + i\sin(n\theta), \quad \text{for all } \theta \in \mathbb{R} \text{ and } n \in \mathbb{Z} \qquad \text{(A.6)}$$

Example A.1 For $\theta \in \mathbb{R}$, let us compute $e^{i(2\theta)} = (e^{i\theta})^2$ in two different ways. Using Euler's Formula directly, we have $e^{i(2\theta)} = \cos(2\theta) + i\sin(2\theta)$. On the other hand, by De Moivre's Theorem, we have:

$$(e^{i\theta})^2 = (\cos\theta + i\sin\theta)^2 = (\cos^2\theta - \sin^2\theta) + i(2\sin\theta\cos\theta)$$

Notice, by equating real and imaginary parts in the two terminal expressions gives the usual double angle identities for cosine and sine.

Fig. A.2 The six 6th roots
of unity, $\zeta_6 = e^{\pi i/3}$,
$\zeta_6^2 = e^{2\pi i/3}$,
$\zeta_6^3 = e^{\pi i} = -1$,
$\zeta_6^4 = e^{4\pi i/3}$, $\zeta_6^5 = e^{5\pi/3}$, and
$\zeta_6^6 = \zeta_6^0 = e^{2\pi} = 1$

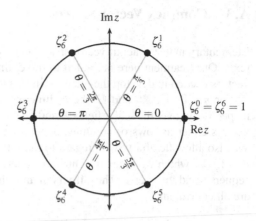

Suppose $n \in \mathbb{N}$. We say that $z \in \mathbb{C}$ is an nth **root of unity** if $z^n = 1$. There are n distinct nth roots of unit, and these occur in the complex plane at equidistant points around the unit circle; see Fig. A.2. Moreover, every nth root of unity is an integer power of $\zeta_n = e^{2\pi i/n}$.

The techniques of calculus carry over to complex functions. In particular, if $f(t) = g(t) + h(t)i$ is a complex function of some parameter $t \in \mathbb{R}$, decomposed in standard form, then $\int f(t)\,dt = \int g(t)\,dt + i \int h(t)\,dt$. Toward our work in Fourier transforms we will need the following important definite integral. Assume $k \in \mathbb{Z}$.

$$\int_{-\pi}^{\pi} e^{ik\theta}\,d\theta = \begin{cases} 0 & \text{if } k \neq 0, \\ 2\pi & \text{if } k = 0 \end{cases} \tag{A.7}$$

In fact because of periodicity, the value of the integral is the same when the bounds are shifted by any real number. More generally, for fixed nonzero $d \in \mathbb{Z}$ and $\theta_0 \in \mathbb{R}$, we have:

$$\int_{\theta_0}^{\theta_0+d} e^{i\frac{2\pi}{d}k\theta}\,d\theta = \begin{cases} 0 & \text{if } k \neq 0, \\ d & \text{if } k = 0 \end{cases} \tag{A.8}$$

The result of (A.7) when $k \neq 0$ may be interpreted geometrically. As θ varies from $-\pi$ to π, the complex values $e^{ik\theta}$ trace out the points of the unit circle with $|k|$ complete passes around the unit circle uniformly in θ. In each pass, the values (points/vectors) encountered and contributing to the integral have opposite value which sum to zero.

There is a related result involving discrete sums of exponentials. As above, assume $k \in \mathbb{Z}$.

$$\sum_{\omega=0}^{d-1} e^{i\frac{2k\pi}{d}\omega} = \begin{cases} 0 & \text{if } k \neq 0, \\ d & \text{if } k = 0 \end{cases} \tag{A.9}$$

A.3   Complex Vector Spaces

Elementary material about vector spaces can be found in [39, 40, 48] among other texts. Our treatment here is limited to the definitions, properties, and operations of vectors needed to develop the material in this text.

For $n \in \mathbb{N}$, let $\mathbb{R}^n$ and $\mathbb{C}^n$ be n-dimensional real and complex vector spaces, respectively (over the scalar fields $\mathbb{R}$ and $\mathbb{C}$, respectively). We may write or visualize vectors either as rows or columns, depending on which form is more convenient. We also allow flexibility with regard to the integer segment $k = k_0, k_0 + 1, \ldots k_0 + n - 1$ with which we index the entries of these vectors. We often write a vector as a sequence and vice versa. Thus, if $\mathbf{v}$ is in an r-dimensional space, then the following are all equivalent:

$$\mathbf{v} = (v_k)_{k=k_0}^{k_0+n-1} = (v_{k_0}, v_{k_0+1}, \ldots, v_{k_0+r-1})$$

If $A, B \neq \emptyset$, then $A^B = \{f : B \to A\}$ is the set of all functions from B to A. In particular, $\mathbb{C}^{\mathbb{Z}}$ is the vector space of all complex-valued functions defined on the integers, i.e., sequences of complex numbers indexed by the integers. Although we may use these infinite vectors, in practice either the vectors under consideration will belong to a finite- dimensional subspace, or all but a finite number of entries will be zero.

Define an **inner product** (or **dot product**) on $\mathbb{C}^n$ in the standard way. That is, for all $u, v \in \mathbb{C}^n$, define:

$$\langle u, v \rangle = \sum_{k=k_0}^{k_0+n-1} u_k \overline{v_k} \tag{A.10}$$

The inner product satisfies the following properties.

Theorem A.4 Let $u, v, w \in \mathbb{C}^n$, and $c \in C$.

(a) $\langle u, u \rangle \geq 0$, with equality if and only if u is the zero vector.

(b) $\langle u, v \rangle = \overline{\langle v, u \rangle}$

(c) $\langle u + v, w \rangle = \langle u, w \rangle + \langle v, w \rangle$

(d) $\langle cu, v \rangle = c \langle u, v \rangle$

Then the **norm** (or **length**, **magnitude**) of $u \in \mathbb{C}^n$ is given by the formula:

$$\|u\| = \langle u, u \rangle^{1/2} = \left(\sum_{k=k_0}^{k_0+n-1} |u_k|^2 \right)^{1/2} \tag{A.11}$$

Formula A.11 is referred to as the **Euclidean norm**. This norm generates the **Euclidean metric** on $\mathbb{C}^n$, in which the distance between vectors v and w is given by the usual distance formula, $\|v - w\| = \sqrt{\sum(|v_j - w_j|^2)}$. We now state a version of the **Pythagorean Theorem** that generalizes the usual Pythagorean Theorem [39, 40] to collections of more than two vectors (see also Exercise A.13).

Theorem A.5 *If $v_1, v_2, \ldots, v_r \in \mathbb{C}^n$ are pairwise orthogonal, then we have:*

$$\sum_{k=1}^{r} ||v_k||^2 = \left|\left|\sum_{k=1}^{r} v_k\right|\right|^2 \tag{A.12}$$

The Euclidean norm is also called the **2-norm**. As such, we may write $||\cdot||_2$, with the subscript emphasizing that we are using the 2-norm. A norm measures the *size* or *length* of a vector, and on occasion we may be interested in different interpretations of size. Two very useful norms are the 1-norm and ∞-norm, defined by the following formulas.

$$||v||_1 = \sum_{k=k_0}^{k_0+n-1} |v_k| \tag{A.13}$$

$$||v||_\infty = \max\{|v_k| : k_0 \le k \le k_0+n-1\} \tag{A.14}$$

The 1-norm simply adds the absolute values of all entries, while the ∞-norm gives the entry with largest absolute value. It is straightforward to show:

$$||v||_\infty \le ||v||_2 \le ||v||_1 \tag{A.15}$$

We say that two vectors $u, v \in \mathbb{C}^n$ are **orthogonal** if $\langle u, v \rangle = 0$. An **orthonormal basis** for $\mathbb{C}^n$ is a set of vectors, $\{u_1, u_2, \ldots, u_n\}$ such that $||u_k|| = 1$ for each k, if $j \ne k$ then u_j and u_k are orthogonal, and every vector $v \in \mathbb{C}^n$ can be written in the form,

$$v = c_1 u_1 + c_2 u_2 + \cdots + c_n v_n, \tag{A.16}$$

for some scalars $c_k \in \mathbb{C}$. An orthonormal basis is particularly useful because the coefficients c_k in (A.16) can be found quite easily by $c_k = \langle u_k, v \rangle$.

A function $T : \mathbb{C}^n \to \mathbb{C}^m$ is called a **linear transformation** if the following two properties hold.

1. $T(u + v) = T(u) + T(v)$, for all $u, v \in \mathbb{C}^n$
2. $T(\lambda u) = \lambda T(u)$, for all $u \in \mathbb{C}^n, \lambda \in \mathbb{C}$.

Suppose now that $n = m$ so that $T : \mathbb{C}^n \to \mathbb{C}^n$. We say that $\lambda \in \mathbb{C}$ is an **eigenvalue** for T associated to an **eigenvector** $v \in \mathbb{C}^n$ if $v \ne 0$ and

$$T(v) = \lambda v$$

The set of all eigenvectors associated to a given eigenvalue λ, together with the zero vector, forms a **subspace** of $\mathbb{C}^n$ called the λ-**eigenspace**.

Exercises

A.1 Let $\theta \in \mathbb{R} \setminus \{0, \pm 2\pi, \pm 4\pi, \ldots\}$. Express the following complex numbers in standard rectangular form:

(a) $\dfrac{2}{1 - e^{i\theta}}$ (b) $\dfrac{1}{1 - e^{i\theta}} - \dfrac{1}{1 - e^{-i\theta}}$

A.2 Prove Euler's Identities (A.5) by solving the algebraic system determined by using Euler's Formula on $e^{i\theta}$ and $e^{-i\theta}$.

A.3 Prove all parts of Theorem A.2.

A.4 Let $z = 1 + i$. Express z in polar form, and find the powers z^2, z^3, and z^4. Plot these points together with z^0 and z^1 on the complex plane and describe the progression using geometric language.

A.5 Rewrite $\displaystyle\sum_{k=1}^{7} \sin\left(\dfrac{k\pi}{8}\right)$ without the summation. *Hint:* Use Euler's Identity for sine.

A.6 Verify (A.7). *Hint:* For $k \neq 0$ rewrite the integrand using Euler's Formula. Then use an appropriate substitution to establish (A.8).

A.7 Prove (A.9) by interpreting the sum as a finite geometric series.

A.8 Evaluate $\displaystyle\int_{-\pi}^{\pi} \cos^{2m}(t)\, dt$.

A.9 Let $I_m = \int_0^{2\pi} \cos(x) \cos(2x) \cdots \cos(mx)\, dx$. Determine the conditions on the natural number m so that $I_m \neq 0$. *(Adapted from problem A5 from the Forty-Sixth William Lowell Putnam Mathematical Competition [34].)*

A.10 Establish the following identity related to the *Dirichlet kernel*. *Hint:* Use Euler's Formula and the formula for the sum of a geometric series.

$$\frac{\sin([2j + 1]t)}{\sin(t)} = \sum_{k=-j}^{j} e^{2kti}$$

A.11 Show that Theorem A.4 implies the following for $u, v, w \in \mathbb{C}^n$ and $c \in \mathbb{C}$.

(a) $\langle u, v + w \rangle = \langle u, v \rangle + \langle u, w \rangle$ (b) $\langle u, cv \rangle = \overline{c}\langle u, v \rangle$

A.12 Let $n \in \mathbb{N}$, and let $v \in \mathbb{C}^n$. Prove the inequality.

$$\frac{1}{\sqrt{n}}||v||_1 \leq ||v||_2 \leq \sqrt{n}||v||_\infty \tag{A.17}$$

A.13 Let $r > 1$. Show that if a vector v_r is orthogonal to each of the vectors $v_1, v_2, \ldots, v_{r-1}$, then $v_r \perp w$, where $w = v_1 + v_2 + \cdots + v_{r-1}$. Use this observation, along with induction to prove the general Pythagorean Theorem (A.12).. You may assume the base case with $r = 2$, i.e., if $v_1 \perp v_2$, then $||v_1||^2 + ||v_2||^2 = ||v_1 + v_2||^2$.

A.14 Suppose $T : \mathbb{C}^n \to \mathbb{C}^n$ is a linear transformation and $\lambda \in \mathbb{C}$ is an eigenvalue of T. Let $U = \{v \in \mathbb{C}^n \mid T(v) = \lambda v\}$, i.e., the λ-eigenspace of T. Prove that U is a subspace of $\mathbb{C}^n$ by establishing the following properties.

(a) $T(u + v) = \lambda(u + v)$ for all $u, v \in U$
(b) $T(\mu u) = \lambda \mu u$ for all $u \in U$ and $\mu \in \mathbb{C}$

Appendix B
Triangular Lattices

In the body of this text, we used discrete Fourier methods to count walks that were essentially confined to intervals or rectangular lattices of arbitrary dimension. In this appendix, we extend those results to counting walks in more general regions – in particular, certain bounded regions that tile the plane via reflections, including *triangular lattices*. Triangular lattices, and related structures known as *wedges*, have been studied by Mortimer-Prellberg [44] among others. Grabiner [29], Gessel-Zeilberger [28], and Gessel-Krattenhaler [27] studied the unbounded cases (Weyl alcoves) using the idea of *reflectable walks*. Our Fourier methods, which also rely on lattice-stabilizing reflections, allow one to explicitly count the number of walks within a bounded triangular lattice with an arbitrary initial point. We present the ideas in a rather informal way, showing how our methods can be used to address a specific set of examples and thereby allowing the reader to explore and extend these results to a wider array of problems. We hope that the material in this appendix may serve as a launchpad to more extended research questions in lattice-path enumeration.

The key technique is to first transform the lattice into an equivalent rectangular lattice so that methods from Chap. 3 may then be used. The main steps are outlined below:

1. Identify the lattice type and region on which to count walks.
2. If the lattice type is related to a reflection group (with appropriate properties as explained below), then use a linear or affine transformation to change the lattice so that the region is a fundamental region of the form C^d, and the transition operator can be written as a sum of shifts.
3. Use discrete Fourier techniques to solve the problem in the rectangular lattice.
4. Interpret solutions in the original lattice.

Remark B.1 In order to fully apply and understand the material of this appendix, it is necessary to have a background in elementary group theory and linear algebra. We recommend Armstrong [5] for group theory, and Dummit and Foote [20] as well for group theory and other areas of abstract algebra, including matrix methods and linear algebra.

© Springer Nature Switzerland AG 2019
S. Ault and C. Kicey, *Counting Lattice Paths Using Fourier Methods*, Applied and Numerical Harmonic Analysis,
https://doi.org/10.1007/978-3-030-26696-7

B.1   Triangular Lattices and Triangle Groups

In what follows, we use the standard notation A_2 to mean the lattice created by tiling the plane $\mathbb{R}^2$ with identical equilateral triangles. The vertices of A_2 can be found as integral linear combinations of two generating vectors; we have fixed generators $\alpha = (1, 0)$ and $\beta = \left(-\frac{1}{2}, \frac{\sqrt{3}}{2}\right)$, as shown in Fig. B.1. Hence, we have:

$$A_2 = \{a\alpha + b\beta \mid a, b \in \mathbb{Z}\}$$

We will demonstrate how our Fourier methods can be modified to count paths in triangular regions of both A_2 and the usual square lattice $\mathbb{Z}^2$. First consider tiling the plane by congruent triangles in such a way so that reflections in lines containing a boundary segment of any triangle leave the tiling invariant. Thus, the tiling is said to be generated by a prototype triangle Ω, called a **Möbius triangle**. The set

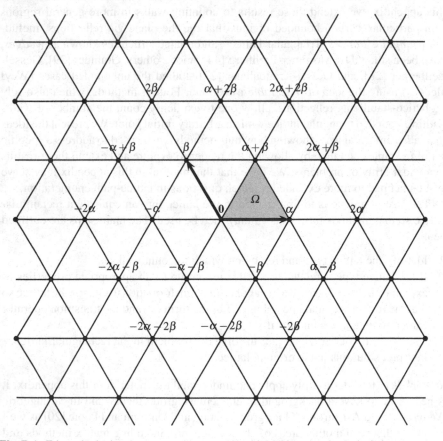

Fig. B.1 Triangular lattice A_2 generated by α and β with Möbius triangle Ω shaded

Fig. B.2 Möbius triangle
$\Omega = \triangle P_1 P_2 P_3$ and
generating reflections for a
triangle group $\Delta(\ell_1, \ell_2, \ell_3)$

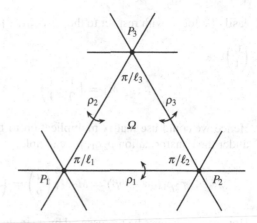

of reflections comprise a **triangle group**[3] Δ. Group composition is handled much
like function composition, in the sense that $\gamma\rho$ stands for first reflecting by ρ, and
then reflecting by γ. Any reflection in a line will reverse orientations; hence an
even number of composed reflections preserves the orientation, and an odd number
reverses it. We say $\rho \in \Delta$ has **even parity** if ρ preserves orientation, and **odd parity**
otherwise. Define a parity function p by:

$$p(\rho) = \begin{cases} 0 & \text{if } \rho \text{ is even,} \\ 1 & \text{if } \rho \text{ is odd} \end{cases} \tag{B.1}$$

The triangle group associated with a triangle with angles π/ℓ_1, π/ℓ_2, π/ℓ_3 is
denoted $\Delta(\ell_1, \ell_2, \ell_3)$. Each triangle group has a standard presentation,

$$\Delta(\ell_1, \ell_2, \ell_3) = \langle \rho_1, \rho_2, \rho_3 \mid \rho_1^2 = \rho_2^2 = \rho_3^2 = (\rho_1\rho_2)^{\ell_1} = (\rho_1\rho_3)^{\ell_2} = (\rho_2\rho_3)^{\ell_3} = 1 \rangle,$$

where ρ_1, ρ_2, and ρ_3 represent reflection in each of the three lines bounding a fixed
Möbius triangle.[4] Figure B.2 shows a Möbius triangle along with angle and reflection
notations consistent with the definition above.

Example B.1 Consider the three generating reflections $\rho_i \in \Delta(3, 3, 3)$ with respect
to the Möbius triangle and two generators, α, β.

- ρ_1: reflection over the horizontal axis. So $\rho_1(\alpha) = \alpha$ and $\rho_1(\beta) = -\alpha - \beta$.
- ρ_2: reflection over line through 0 and $\alpha + \beta$. So $\rho_2(\alpha) = \beta$ and $\rho_2(\beta) = \alpha$.
- ρ_3: reflection over line through α and $\alpha + \beta$. So $\rho_3(\alpha) = \alpha$ and $\rho_3(\beta) = 2\alpha + 2\beta$.

The reflections ρ_1 and ρ_2 are linear with respect to the underlying vector space $\mathbb{R}^2$,
while ρ_3 is an affine transformation. Representing matrices M_i for ρ_1 and ρ_2 can

---

[3]See [16, 32, 41] for background and theory concerning reflection groups. Only the most basic
knowledge of such groups is needed for our treatment.
[4]This presentation implies that a triangle group is a *Coxeter group*.

easily be found with respect to the basis $\{\alpha, \beta\}$, identifying α with $\begin{pmatrix} 1 \\ 0 \end{pmatrix}$ and β with $\begin{pmatrix} 0 \\ 1 \end{pmatrix}$.

$$M_1 = \begin{pmatrix} 1 & -1 \\ 0 & -1 \end{pmatrix}, \qquad M_2 = \begin{pmatrix} 0 & 1 \\ 1 & 0 \end{pmatrix}$$

Hence we could use matrix multiplication to find the image of any point $a\alpha + b\beta$ under the transformation $\rho_2\rho_1$, for example.

$$\rho_2\rho_1(a\alpha + b\beta) = M_2 M_1 \begin{pmatrix} a \\ b \end{pmatrix} = \begin{pmatrix} 0 & -1 \\ 1 & -1 \end{pmatrix} \begin{pmatrix} a \\ b \end{pmatrix} = \begin{pmatrix} -b \\ a - b \end{pmatrix}$$

However, ρ_3 cannot be represented by a single matrix; it is an affine (but not linear) transformation. Moreover, the form of the transformation depends on the size of the Möbius triangle Ω. If Ω has vertices at $\mathbf{0}$, $d\alpha$, and $d\alpha + d\beta$ (as Fig. B.5 illustrates for $d = 6$), then we have:

$$\rho_3 \begin{pmatrix} a \\ b \end{pmatrix} = \begin{pmatrix} -1 & 0 \\ -1 & 1 \end{pmatrix} \begin{pmatrix} a \\ b \end{pmatrix} + \begin{pmatrix} 2d \\ d \end{pmatrix}$$

The form of this equation comes from the observations that $\rho_3(\mathbf{0}) = 2d\alpha + d\beta$, $\rho_3(\alpha) = (2d - 1)\alpha + (d - 1)\beta$, and $\rho_3(\beta) = 2d\alpha + (d + 1)\beta$.

There are only three possible triangle groups in the plane [41], up to permutation of the parameters ℓ_i.

- $\Delta(3, 3, 3)$, associated with an equilateral triangle.
- $\Delta(2, 3, 6)$, associated with a 30°–60°–90° triangle.
- $\Delta(2, 4, 4)$, associated with an isosceles right triangle.

The lattice A_2 is left invariant under the action of $\Delta(3, 3, 3)$ and $\Delta(2, 3, 6)$, while $\mathbb{Z}^2$ is left invariant under the action of $\Delta(2, 4, 4)$ (see Fig. B.3). We take the convention that the parameters ℓ_1, ℓ_2, ℓ_3 are given in order of the angles π/ℓ_i as the triangle is traversed counter-clockwise starting at the origin. So while $\Delta(4, 4, 2) \cong \Delta(4, 2, 4)$ as groups, we distinguish them because the orientations of the Möbius triangles within the lattice are distinct.

B.2 Tilings of the Plane and Corridors

Let $d \in \mathbb{N}$ and consider a Möbius triangle Ω with vertices P_1 at the origin, P_2 located on the horizontal axis, and P_3 situated in the first quadrant of the lattice, as suggested by Fig. B.2. By definition, the associated triangle group Δ exhibits a **group action** on the lattice sending Ω to various copies of itself within the tiling.

Fig. B.3 Isoceles right triangles tiling the plane, with corresponding triangle group $\Delta(2, 4, 4)$. The vertices coincide with the $\mathbb{Z}^2$ lattice

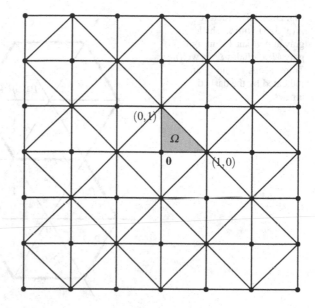

Table B.1 Three triangle group structures in two dimensional lattices. The parameter d controls the size of Ω, and the restriction ensures a nontrivial corridor

Triangle group (Δ)	Lattice	Vertices P_1, P_2, P_3	Restriction on d
$\Delta(3, 3, 3)$	A_2	$\mathbf{0}$, $d\alpha$, $d\alpha + d\beta$	$d \geq 3$
$\Delta(4, 4, 2)$	$\mathbb{Z}^2$	$\mathbf{0}$, $(d, 0)$, $\left(\frac{d}{2}, \frac{d}{2}\right)$	$d \geq 4$
$\Delta(4, 2, 4)$	$\mathbb{Z}^2$	$\mathbf{0}$, $(d, 0)$, (d, d)	$d \geq 3$

For each $\gamma \in \Delta$, let $\gamma\Omega$ be the image of Ω under the transformation γ. Then the set of triangles $\{\gamma\Omega \mid \gamma \in \Delta\}$ tiles the plane. For economy of notation we typically label the triangle $\gamma\Omega$ simply by γ, especially in figures.

It turns out that the six possible tilings associated with the group $\Delta(2, 3, 6)$ fail to admit the peculiar structure needed for our Fourier methods to work (briefly, the associated *fundamental regions* are too close together—more about this later). Furthermore, the tiling for $\Delta(2, 4, 4)$ is equivalent to that of $\Delta(4, 2, 4)$ (in the sense that they determine equivalent regions of their lattice). Thus only $\Delta(3, 3, 3)$, $\Delta(4, 4, 2)$, and $\Delta(4, 2, 4)$ can be analyzed using our methods. Table B.1 summarizes the triangle groups under consideration in this chapter.

By properties of triangle groups, there exist elements τ_1, $\tau_2 \in \Delta$ representing two independent translations of the plane such that all translations that preserve the tiling are generated by these two (indeed many choices for τ_1, τ_2 exist). The translation subgroup $\Sigma = \langle \tau_1, \tau_2 \rangle$ is normal in Δ (which is easily verified: *reflecting* followed by *translating* followed by *reflecting back* simply results in another translation); let $\overline{\Delta} = \Delta/\Sigma$ be the quotient group. For each $\overline{\rho} \in \overline{\Delta}$, choose a representative $\rho \in \Delta$ (the choice is immaterial). Let Ψ be the region of the plane, which we call the

Fig. B.4 Tile structure $\Delta(3, 3, 3)$ with Ω shaded gray and Ψ shaded in light gray (Ψ includes Ω as well). Translations τ_1 and τ_2 are indicated by the dashed arrows

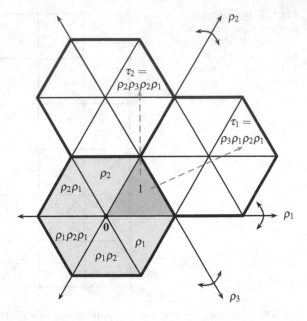

Table B.2 Fundamental tiles and generating translations for each structure mentioned in Table B.1

Triangle group (Δ)	Fundamental tile (Ψ)	τ_1	τ_2
$\Delta(3, 3, 3)$	Hexagon (6 triangles)	$\rho_3\rho_1\rho_2\rho_1$	$\rho_2\rho_3\rho_2\rho_1$
$\Delta(4, 4, 2)$	Square (8 triangles)	$\rho_3\rho_1\rho_2\rho_1$	$\rho_2\rho_1\rho_3\rho_1$
$\Delta(4, 2, 4)$	Square (8 triangles)	$\rho_3\rho_2\rho_1\rho_2$	$\rho_2\rho_3\rho_2\rho_1$

fundamental tile, consisting of the triangles $\rho\Omega$ as $\overline{\rho}$ ranges in $\overline{\Delta}$. By construction, Ψ must tile the plane via translation only. The choices of representative elements ρ may be arbitrary, however convenient choices may be made so that Ψ is connected, contains Ω, and is centered at the origin, as illustrated in Fig. B.4 for $\Delta(3, 3, 3)$. Table B.2 gives a particular choice of translations and fundamental tiles for the three triangle groups that we have been considering thus far.

The set of lattice points (in A_2 or $\mathbb{Z}^2$, as the case may be) that are interior to Ω (i.e., not including the boundary) will be called the **fundamental region**, C^Ω, in direct analogy with the fundamental region as in Definitions 1.1 or 3.2. We are interested in counting corridor paths in Ω-**corridor**, $\mathbb{N}_0 \times C^\Omega$, or equivalently, walks in a graph G_Ω whose vertices are the points of C^Ω and whose edges correspond to the allowable moves in the lattice.[5] Toward this end, we found it necessary to consider the reflected regions, ρC^Ω, as something like the *dual corridor* structure of Sect. 2.1. If $p(\rho) = 0$ (respectively, $p(\rho) = 1$), then the set ρC^Ω is called a *positive*

---

[5]Recall $G_{\mathbf{d}, \mathscr{M}}$ from Sect. 3.1.

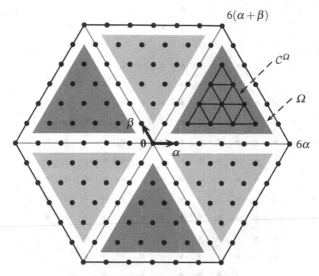

Fig. B.5 Fundamental tile Ψ for $\Delta(3, 3, 3)$ with $d = 6$, in A_2. Fundamental region $C^{\Omega} \subseteq \Omega$ shown with allowable moves. Positive regions in gray; negative regions in light gray

(respectively, *negative*) region. Figures B.5, B.4 and B.7 show the tile and corridor structures associated with the triangle groups shown in Tables B.1 and B.2.

Assume now that we have chosen a reflection group Δ and tile structure (e.g., from Table B.1), and suppose Λ is a lattice (either $\mathbb{Z}^2$ or A_2) that is invariant under Δ. For each $\gamma \in \Delta$ define an operator H^{γ} on functions $v : \Lambda \to \mathbb{Z}$, similar to that defined by Eq. (3.6).

$$H^{\gamma}[v](\mathbf{x}) = v(\gamma^{-1}\mathbf{x}), \quad \forall \mathbf{x} \in \Lambda \tag{B.2}$$

These operators compose in the usual way: for $\eta, \gamma \in \Delta$, we have $H^{\eta} H^{\gamma} = H^{\eta\gamma}$. Extending Definition 3.5, we define *admissible states* in the context of a given reflection group.

Definition B.1 We call a function $v : \Lambda \to \mathbb{Z}$ an **admissible state** with respect to a reflection group Δ if for every $\gamma \in \Delta$, we have $H^{\gamma}[v] = (-1)^{p(\gamma)}v$.

When $\rho \in \Delta$ is a reflection in the boundary line of the Möbius triangle containing the fundamental region, the condition $H^{\rho}[v] = -v$ corresponds to the *antisymmetry* condition in the dual corridor construction. Definition B.1 also captures periodicity and the "zero value on boundary" conditions. Since every translation τ has even parity, we have $v(\tau^{-1}\mathbf{x}) = H^{\tau}[v](\mathbf{x}) = v(\mathbf{x})$; hence with τ_1 and τ_2 chosen as in Table B.2, an admissible state must be periodic with independent periods τ_1, τ_2. Moreover, if $\mathbf{x}$ is on any boundary line, and $\gamma \in \Delta$ corresponds to reflection over that line (such γ must exist by properties of reflection groups), then $v(\mathbf{x}) = 0$ (see Exercise B.7).

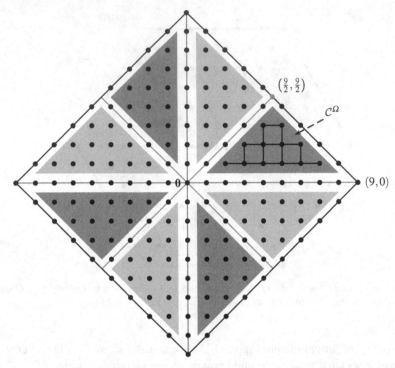

Fig. B.6 Fundamental tile Ψ for $\Delta(4, 4, 2)$ with $d = 9$, in Z^2. Fundamental region $C^\Omega \subseteq \Omega$ shown with allowable moves. Positive regions in gray; negative regions in light gray

Let us now define the operators that will serve as transitions from state to state. For each point $\mathbf{m} \in \Lambda$, define a *shift operator* $R^{\mathbf{m}}$ analogous to Eq. (3.7) by the following rule.

$$R^{\mathbf{m}}[\nu](\mathbf{x}) = \nu(\mathbf{x} - \mathbf{m}), \quad \text{for all } \mathbf{x} \in \Lambda \tag{B.3}$$

We note that when $\tau \in \Delta$ is a translation, then $R^{\mathbf{m}} = H^\tau$ for $\mathbf{m} = \tau(\mathbf{0})$. As before, a **move** is a vector whose components are each in $\{-1, 0, 1\}$. For the kinds of regions we study in this section, the allowable nonidentity moves in $\mathbb{Z}^2$ will be restricted to those of the form $(\pm 1, 0)$ and $(0, \pm 1)$, while the allowable nonidentity moves in A_2 correspond to the six directions $\{\pm\alpha, \pm\beta, \pm(\alpha + \beta)\}$. Suppose $\mathcal{M}$ is a subset of allowable moves, and let $T = \sum_{\mathbf{m} \in \mathcal{M}} R^{\mathbf{m}}$ be a transition operator. In order for our Fourier methods to work, we require T to preserve admissible states (compare Definition 3.7).

Definition B.2 An operator T is called **balanced** if T is Δ-invariant. That is, for every $\gamma \in \Delta$, we have $H^\gamma T = T H^\gamma$.

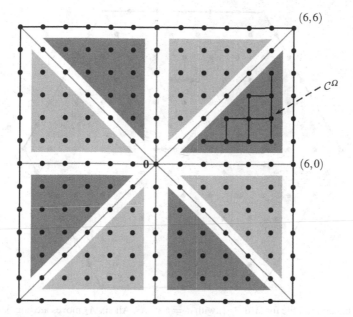

Fig. B.7 Fundamental tile Ψ for $\Delta(4, 2, 4)$ with $d = 6$, in Z^2. Fundamental region $C^\Omega \subseteq \Omega$ shown with allowable moves. Positive regions in gray; negative regions in light gray

Once a transition operator $T = \sum_{\mathbf{m} \in \mathcal{M}} R^{\mathbf{m}}$ has been identified, the set of moves $\mathcal{M}$ may then be called balanced.[6] The main idea is that balanced moves will "cancel" each other out. Just as in the rectangular dual corridor structure, any move into the "no-man's land" between positive and negative regions should be negated by a move coming from the adjacent negative region. In this way, a seemingly complicated *reflection principle* is encoded directly into the structure of the state functions and balanced transitions that act upon them. Moreover, balanced transition operators send admissible states to admissible states.

Lemma B.1 *A balanced operator T sends admissible states to admissible states.*

Proof Suppose v is admissible; we must show $T[v]$ also is. Let $\gamma \in \Delta$ be arbitrary. We have $H^\gamma T[v] = T H^\gamma[v] = (-1)^{p(\gamma)} T[v]$, showing that $T[v]$ satisfies Definition B.1. $\qquad\square$

On the other hand, we encounter a new obstacle in this more general setting. When the fundamental region is triangular, there is the possibility that a move might take us directly from a positive to a negative region, bypassing "no-man's land" altogether. In order to prevent this, we also require our moves to be **local**, in the sense that a path that starts in the fundamental region should not be permitted to move into another

---

[6]It can easily be verified that Definition B.2 is equivalent to Definition 3.7 when Δ is the reflection group generated by reflections ρ_j and σ_j $(j = 1, 2)$ as defined in Sect. 3.2.

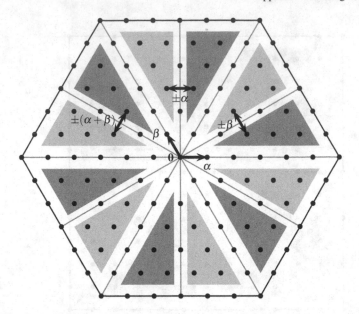

Fig. B.8 Fundamental tile for $\Delta(6, 3, 2)$ with $d = 6$ in A_2. All six A_2 moves are nonlocal

region. All six A_2 moves are local in the $\Delta(3, 3, 3)$ tile structure. In the $\Delta(4, 4, 2)$ and $\Delta(4, 2, 4)$ structures, the only local moves are in the cardinal directions: $(\pm 1, 0)$, $(0, \pm 1)$ (see Figs. B.6, B.8 and B.7). All six of the A_2 moves are nonlocal with respect to the $\Delta(6, 3, 2)$ tiling (see Fig. B.8).

In the lattice A_2, there are two balanced transition operators with respect to $\Delta(3, 3, 3)$ consisting of three moves:

- $T^+ = R^\alpha + R^\beta + R^{-\alpha - \beta}$
- $T^- = R^{-\alpha} + R^{-\beta} + R^{\alpha + \beta}$
- The sum, $T_{A_2} = T^+ + T^-$

Let's check that T^+ is balanced. Definition B.2 is equivalent to checking that the set of moves $\mathcal{M}^+ = \{\alpha, \beta, -\alpha - \beta\}$ is invariant under the two reflections ρ_1 and ρ_2 (compare Definition 3.7).

$$\rho_1 \mathcal{M}^+ = \{\rho_1(\alpha), \rho_1(\beta), \rho_1(-\alpha - \beta)\} = \{\alpha, -\alpha - \beta, -\beta\} = \mathcal{M}^+$$

$$\rho_2 \mathcal{M}^+ = \{\rho_2(\alpha), \rho_2(\beta), \rho_2(-\alpha - \beta)\} = \{\beta, \alpha, -\alpha - \beta\} = \mathcal{M}^+$$

Because of the diagonal reflection in $\Delta(4, 4, 2)$ or $\Delta(4, 2, 4)$, the only nonidentity balanced transition in this context is $T = R^{(1,0)} + R^{(-1,0)} + R^{(0,1)} + R^{(0,-1)} = T_1 + T_2$.

The upshot is that whenever v_0 is an admissible state encoding the initial point $\mathbf{a} \in \mathcal{C}^\Omega$, and if $\mathcal{M}$ is a set of local moves such that $T = \sum_{\mathbf{m} \in \mathcal{M}} R^{\mathbf{m}}$ is balanced, then

Table B.3 Coordinates of $\mathbf{x}_i$ corresponding to generating translations, invariant factors d_1, d_2, and transformation matrix P. Parameter d controls the size of the corridor, as in Table B.1

Triangle group (Δ)	$\mathbf{x}_1 = \tau_1(\mathbf{0})$	$\mathbf{x}_2 = \tau_2(\mathbf{0})$	(d_1, d_2)	P	A	Q	A_{norm}
$\Delta(3,3,3)$	$2d\alpha + d\beta$	$d\alpha + 2d\beta$	$(d, 3d)$	$\begin{pmatrix} 0 & 1 \\ 1 & -2 \end{pmatrix}$	$\begin{pmatrix} 2d & d \\ d & 2d \end{pmatrix}$	$\begin{pmatrix} 1 & 2 \\ 0 & -1 \end{pmatrix}$	$\begin{pmatrix} d & 0 \\ 0 & 3d \end{pmatrix}$
$\Delta(4,4,2)$	(d, d)	$(-d, d)$	$(d, 2d)$	$\begin{pmatrix} 0 & 1 \\ 1 & -1 \end{pmatrix}$	$\begin{pmatrix} 1 & -1 \\ 1 & 1 \end{pmatrix}$	$\begin{pmatrix} 1 & 1 \\ 0 & -1 \end{pmatrix}$	$\begin{pmatrix} d & 0 \\ 0 & 2d \end{pmatrix}$
$\Delta(4,2,4)$	$(2d, 0)$	$(0, 2d)$	$(2d, 2d)$	$\begin{pmatrix} 1 & 0 \\ 0 & 1 \end{pmatrix}$	$\begin{pmatrix} 2d & 0 \\ 0 & 2d \end{pmatrix}$	$\begin{pmatrix} 1 & 0 \\ 0 & 1 \end{pmatrix}$	$\begin{pmatrix} 2d & 0 \\ 0 & 2d \end{pmatrix}$

$v_n = T^n[v_0]$ will correctly count the number of n-length paths staying in the corridor $\mathbb{N}_0 \times \mathcal{C}^{\Omega}$, beginning at $(0, \mathbf{a})$, ending at $(n, \mathbf{x})$, allowing only moves from $\mathcal{M}$.

Now that we have defined admissible states and balanced transitions, we are in good position to use our Fourier methods to produce an explicit formula for the vertex state function $v_n(\mathbf{x}) = T^n[v_0](\mathbf{x})$, but in order to do this, the lattice must be transformed. A lattice Λ is an Abelian group under addition, so when there is a tiling of a plane lattice Λ by a region Ψ via translation by independent vectors $\mathbf{x}_1, \mathbf{x}_2 \in \Lambda$, then the quotient $\Psi = \Lambda/\langle \mathbf{x}_1, \mathbf{x}_2 \rangle$ is a finite abelian group. The Fundamental Theorem of Abelian Groups implies that there is an isomorphism $f : \Psi \cong \mathbb{Z}/d_1\mathbb{Z} \times \mathbb{Z}/d_2\mathbb{Z}$ for some integers d_1, d_2. Such an isomorphism can be found by *Invariant Factor Decomposition* (See [50] Chap. 6, and especially Theorem 6.15.), using methods from linear algebra such as row and column reduction of matrices.

In our situations, we have identified generating translations $\tau_i \in \Lambda$, and so $\mathbf{x}_i = \tau_i(\mathbf{0})$ $(i = 1, 2)$. The method used to transform the lattice requires finding the *Smith Normal form (SNF)* of the matrix A whose columns are $\mathbf{x}_1$ and $\mathbf{x}_2$, obtaining a diagonal matrix A_{norm} along with tranformation matrices P and Q such that $A_{norm} = PAQ$. The matrix P may be used to represent the isomorphism f mentioned above (see the commutative diagram (B.4), which illustrates how P represents f), and d_i are the diagonal elements of A_{norm}. The methods for finding an SNF are elementary, and so for convenience, the results for each corridor structure are provided in Table B.3). Indeed many computer mathematics systems, including sage, are capable of finding the SNF decomposition of matrices.

$$
\begin{array}{ccccccc}
0 & \longrightarrow & \mathbb{Z}^2 & \xrightarrow{A} & \mathbb{Z}^2 & \longrightarrow & \mathbb{Z}^2/A\mathbb{Z}^2 \cong \Psi & \longrightarrow & 0 \\
& & \downarrow{Q} & & \downarrow{P} & & \downarrow{f} & & \\
0 & \longrightarrow & \mathbb{Z}^2 & \xrightarrow{A_{norm}} & \mathbb{Z}^2 & \longrightarrow & \mathbb{Z}^2/d_1\mathbb{Z}^2 \times \mathbb{Z}^2/d_2\mathbb{Z}^2 & \longrightarrow & 0
\end{array}
\tag{B.4}
$$

The following theorem sums up the discussion.

Theorem B.1 *Suppose C^Ω is a fundamental region of a plane lattice, in which Ω is a Möbius triangle having a triangle group shown in Table B.1. The number of n-length paths starting at $\mathbf{a} \in C^\Omega$ and ending at $\mathbf{x} \in C^\Omega$, allowing moves from a balanced set of local moves $\mathscr{M}$, is given by $v_n(P\mathbf{x})$, where*

$$v_n = \mathcal{F}^{-1}\left[\widehat{T}^n V_0\right],$$

where

$$\widehat{T}(\omega_1, \omega_2) = \sum_{\mathbf{m} \in \mathscr{M}} e^{-2\pi i P\mathbf{m}\cdot(\omega_1/d_1, \omega_2/d_2)}, \tag{B.5}$$

where d_i, and P are given in Table B.3, and where V_0 is defined by the following formula.

$$V_0(\omega_1, \omega_2) = \sum_{\overline{\rho} \in \overline{\Delta}} (-1)^{p(\rho)} e^{-2\pi i (P\rho\mathbf{a})\cdot(\omega_1/d_1, \omega_2/d_2)} \tag{B.6}$$

B.3   Application: Walks in Low-Dimensional Wedges

Let $r \geq 1$ be an integer, and for $0 \leq j \leq r$ let $\mathbf{e}_j$ be the unit vector in the x_j-direction in $\mathbb{R}^{r+1}$. The **simplicial lattice**, or **wedge**, of order $N \in \mathbb{N}$ in $\mathbb{R}^{r+1}$ is defined as the set of integral points on the hyperplane $x_0 + x_1 + \cdots + x_r = N$, with edges parallel to the vectors $\mathbf{e}_i - \mathbf{e}_j$, for $0 \leq i < j \leq r$. Taking into account that there are two ways to traverse any given edge, there are a total of $r(r+1)$ allowable moves in the wedge. Walks in the wedge may visit the same vertex more than once, and even backtrack along the same edge multiple times.

Remark B.2 The points of the wedge lie on an *affine* lattice, however we will use linear algebra to interpret the points as vectors in an appropriate vector space.

Example B.2 For $r = 1$, the wedge of order N is the set of lattice points on the line $x_0 + x_1 = N$. There are two moves, corresponding to the vector $\mathbf{e}_0 - \mathbf{e}_1 = \langle 1, -1 \rangle$ and its opposite. Counting walks in this lattice is equivalent to counting walks in the path graph P_{N+1}, or counting corridor paths in $C^{(N+2)}$, as demonstrated in Fig. B.9. A 2-dimensional wedge is illustrated by Fig. B.10.

Unfortunately, wedges of dimension $r \geq 3$ do not admit the kind of reflection invariance that we require for our Fourier methods to be set up, so we must stick to the $r = 2$ wedges. Such two-dimensional structures are equivalent to bounded triangular regions of the triangular lattice A_2.

Consider the triangular wedge defined by $x_0 + x_1 + x_2 = N$, where $N \geq 0$, restricted to points $(x_0, x_1, x_2) \in (\mathbb{N}_0)^3$ (see Fig. B.10). There are six valid moves.

$$\{(1, -1, 0), (-1, 1, 0), (1, 0, -1), (-1, 0, 1), (0, 1, -1), (0, -1, 1)\} \tag{B.7}$$

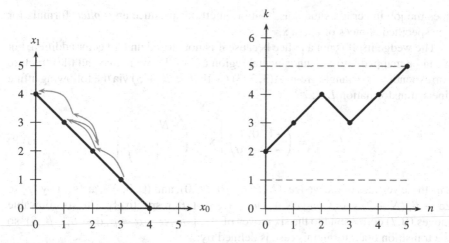

Fig. B.9 Walks in 1-dimensional wedge ($N = 4$ case shown on left; initial point: (3, 1)) correspond to corridor paths ($d = 6$ shown on right; initial point: $k = 2$)

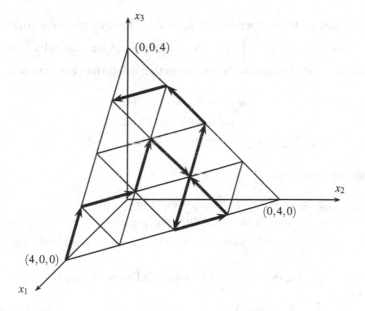

Fig. B.10 A path of length 10 in the $N = 4$ wedge from $\mathbf{u} = (4, 0, 0)$ to $\mathbf{v} = (1, 0, 3)$

We may ask how many n-length walks exist in the wedge starting at a given point $\mathbf{u}$ and ending at $\mathbf{x}$. The problem of counting walks in wedges has been extensively studied; for example, Mortimer and Prellberg [44] develop a generating function that does the job in certain situations,[7] but our methods produce an *explicit* formula for *any* specified choices of **u** and **x**.

The wedge itself is not a lattice because it is not closed under vector addition, but it can be mapped into the fundamental region C^{Ω} of the A_2 lattice and tile structure corresponding to triangle group $\Delta(3, 3, 3)$ (with $d = N + 3$) via the following affine linear transformation $L_N : \mathbb{Z}^3 \to \mathbb{Z}^2$.

$$L_N(\mathbf{x}) = \begin{pmatrix} -1 & 0 & 0 \\ -1 & -1 & 0 \end{pmatrix} \left[\mathbf{x} + \begin{pmatrix} -N-2 \\ 1 \\ 1 \end{pmatrix} \right] \tag{B.8}$$

The three vertices of the wedge, $(N, 0, 0)$, $(0, N, 0)$, and $(0, 0, N)$, are sent by L_N to $2\alpha + \beta$, $(N + 2)\alpha + \beta$, and $(N + 2)\alpha + (N + 1)\beta$, respectively. The six allowable moves (B.7) correspond to the six moves of A_2, $\{-\alpha, \alpha, -\alpha - \beta, \alpha + \beta, -\beta, \beta\}$, so the transition operator in this case is defined by:

$$T_{A_2} = R^{-\alpha} + R^{\alpha} + R^{-\beta} + R^{\beta} + R^{-\alpha-\beta} + R^{\alpha+\beta} \tag{B.9}$$

Using Table B.3, we find the periodicity, $d_1 = N + 3$ and $d_2 = 3(N + 3)$, and transformation matrix $P = \begin{pmatrix} 0 & 1 \\ 1 & -2 \end{pmatrix}$ that serves to transform our triangular lattice into a rectangular one. Let us start with the six moves that will go into our transition matrix.

$$\pm\alpha \mapsto \pm P\alpha = \pm(0, 1)$$
$$\pm\beta \mapsto \pm P\beta = \pm(1, -2)$$
$$\pm(\alpha + \beta) \mapsto \pm P(\alpha + \beta) = \pm(1, -1)$$

Thus $\widehat{T}$ can be generated:

$$\widehat{T}(\omega_1, \omega_2) = e^{-2\pi i \omega_2/(3(N+3))} + e^{2\pi i \omega_2/(3(N+3))}$$
$$+ e^{-2\pi i [\omega_1/(N+3) - 2\omega_2/(3(N+3))]} + e^{2\pi i [\omega_1/(N+3) - 2\omega_2/(3(N+3))]}$$
$$+ e^{-2\pi i [\omega_1/(N+3) - \omega_2/(3(N+3))]} + e^{2\pi i [\omega_1/(N+3) - \omega_2/(3(N+3))]}$$

Using Euler's formula, this result can be expressed more concisely.

$$\widehat{T}(\omega_1, \omega_2) = 2 \left[\cos\left(\frac{2\pi\omega_2}{3[N+3]} \right) + \cos\left(\frac{2\pi[3\omega_1 - 2\omega_2]}{3[N+3]} \right) + \cos\left(\frac{2\pi[3\omega_1 - \omega_2]}{3[N+3]} \right) \right] \tag{B.10}$$

Now let us construct the DFT of the initial state, i.e., V_0. Let $\mathbf{a} = L_N(\mathbf{u})$ in (B.6). As an example, we do this for $\mathbf{u} = (N, 0, 0)$, so $\mathbf{a} = L_N((N, 0, 0)) = 2\alpha + \beta$.

---

[7]For example, in [44] Theorem 3 counts the total n-length paths starting at **u**, but with unspecified terminal point, and Proposition 5 counts walks starting in the center of the lattice and ending on a fixed edge.

ρ	$\rho\mathbf{a}$	$P\rho\mathbf{a}$	$\|\rho\|$
1	$2\alpha + \beta$	$(1, 0)$	$+1$
ρ_1	$\alpha - \beta$	$(-1, 3)$	-1
ρ_2	$\alpha + 2\beta$	$(2, -3)$	-1
$\rho_1\rho_2$	$-\alpha - 2\beta$	$(-2, 3)$	$+1$
$\rho_2\rho_1$	$-\alpha + \beta$	$(1, -3)$	$+1$
$\rho_1\rho_2\rho_1$	$-2\alpha - \beta$	$(-1, 0)$	-1

Thus we have (again using Euler's Formula):

$$V_0 = -2i \left[\sin\left(\frac{2\pi\omega_1}{N+3}\right) + \sin\left(\frac{2\pi[\omega_1 - \omega_2]}{N+3}\right) - \sin\left(\frac{2\pi[2\omega_1 - \omega_2]}{N+3}\right) \right]$$

$$(B.11)$$

Finally, appealing to Theorem B.1, the state function $v_n(PL_N(\mathbf{x}))$ counts the number of n-length paths to a particular point $\mathbf{x}$ in the original triangular lattice, where $v_n = \mathcal{F}^{-1}[\widehat{T}^n V_0]$ with $\widehat{T}$ as in (B.10) and V_0 as in (B.11). It would take up too much space on this page to express the final form of v_n; however, the formula easily can be coded in a computer algebra system.

Example B.3 Consider $N = 2$ and paths beginning at $(2, 0, 0)$ in the wedge. The corridor state progression for $n \le 5$ is shown in Fig. B.11.

The number of paths beginning and ending at $(2, 0, 0)$ is a known sequence,[8] $(1, 0, 2, 2, 12, 30, 110, 336, \ldots)$. We found that the sequence of sums of vertex numbers (i.e., the corridor numbers for the wedge) matches an interesting sequence as well. This sequence, $(1, 2, 8, 24, 80, 256, 832, 2588, 8704, \ldots)$ has general term equal to $2^n F_{n+1}$, where $(F_n)_{n\in\mathbb{N}_0} = (1, 1, 2, 3, 5, \ldots)$ is the sequence of Fibonacci numbers.[9]

Exercises

B.1 Find the number of length 5 walks from $(4, 0, 0)$ to $(0, 0, 4)$ in the wedge shown in Fig. B.10 by carefully listing each walk explicitly. Do the same for walks of length 6.

B.2 Determine the Cartesian coordinates of the following points in A_2. In part (d), assume $d \in \mathbb{Z}$.
(a) $\alpha + \beta$ (b) $-2\alpha + 5\beta$ (c) -24β (d) $d\alpha + d\beta$

B.3 Find the reflection matrices M_1, M_2 corresponding to $\rho_1, \rho_2 \in \Delta(4, 4, 2)$. Then find the affine transformation representing $\rho_3 \in \Delta(4, 4, 2)$. Do the same for $\rho_k \in \Delta(4, 2, 4)$.

B.4 Show the following relations in $\Delta(3, 3, 3)$ geometrically (i.e., by showing how the fundamental triangle Ω transforms).
(a) $\rho_1\rho_2\rho_1 = \rho_2\rho_1\rho_2$ (b) $\rho_1\rho_3\rho_1 = \rho_3\rho_1\rho_3$ (c) $\rho_2\rho_3\rho_2 = \rho_3\rho_2\rho_3$

---

[8] See OEIS sequence A093044.
[9] See OEIS sequences A063727 and A085449.

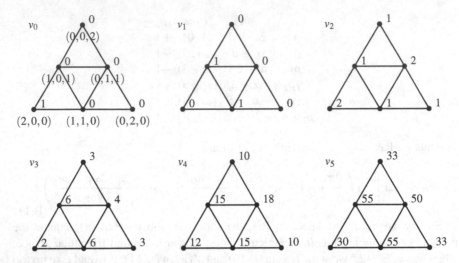

Fig. B.11 Corridor state progression in the $N = 2$ wedge with initial point $\mathbf{u} = (2, 0, 0)$

B.5 Figure B.6 shows eight reflections of the Möbius triangle corresponding to the reflection group $\Delta = \Delta(4, 4, 2)$. Use the diagram to show that the subgroup of Δ generated by ρ_1 and ρ_2 has order eight. Label the eight triangles $\gamma \Omega$ by their appropriate element γ of this subgroup.

B.6 The tiling of the plane given by reflections of a $30°$–$60°$–$90°$ triangle may be regarded as a *subdivision* of the standard A_2 lattice. Starting with triangular graph paper, draw in extra lines subdividing each equilateral triangle to show how the reflected triangles $\gamma \Omega$, for $\gamma \in \Delta(6, 3, 2)$, tile the plane. Outline a fundamental tile and its translates (see Fig. B.8).

B.7 Suppose $\gamma \in \Delta$ is reflection in a line ℓ and v is a Δ-admissible function. Show that $v(\mathbf{x}) = 0$ if $\mathbf{x}$ lies on ℓ.

B.8 Set up the state function v_n for the following triangular corridors in $\mathbb{Z}^2$ with the standard transition operator $T = T_1 + T_2$. Use a computer to work out the state progressions for a few small n values.

(a) Triangle group: $\Delta(4, 4, 2)$ with $d = 9$ (see Fig. B.6)
(b) Triangle group: $\Delta(4, 2, 4)$ with $d = 6$ (see Fig. B.7)

B.9 Verify that the matrices given in Table B.3 are correct by showing $A_{norm} = PAQ$ in each case.

B.10 Prove Theorem B.1.

Research Questions

B.11 Define the concept of random walks in the lattice A_2. How can the methods developed in this chapter be used to study these kinds of random walks?

B.12 Code the results of Theorem B.1 in a mathematical programming language. Use your program to explore state progressions in a variety of triangular corridors. What patterns (if any) emerge?

B.13 Can Theorem B.1 be extended to include unbounded versions of C^Ω (recall Sect. 2.5)?

B.14 Explore cases in which the boundary segments of C^Ω are identified in various ways (recall Sect. 3.4).

B.15 Combine the eigenvector/eigenvalue analysis from Chap. 4 with triangular corridors. See if you can come up with decompositions analogous to those in Theorem 4.5. Determine the dominant eigenvalue(s) and use them to produce approximation formulae.

Selected Solutions

1.1

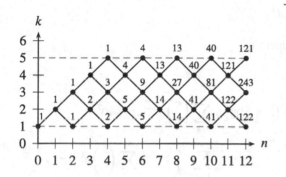

n	State v_n	c_n
0	$v_0=(1,0,0,0,0)$	1
1	$v_1=(0,1,0,0,0)$	1
2	$v_2=(1,0,1,0,0)$	2
3	$v_3=(0,2,0,1,0)$	3
4	$v_4=(2,0,3,0,1)$	6
5	$v_5=(0,5,0,4,0)$	9
6	$v_6=(5,0,9,0,4)$	18
7	$v_7=(0,14,0,13,0)$	27
8	$v_8=(14,0,27,0,13)$	54
9	$v_9=(0,41,0,40,0)$	81
10	$v_{10}=(41,0,81,0,40)$	162
11	$v_{11}=(0,122,0,121,0)$	243
12	$v_{12}=(122,0,243,0,121)$	486

1.2

n	$v_{n,2}^{(4)}$	$v_{n,3}^{(4)}$	$v_{n,4}^{(4)}$
0	(0, 1, 0, 0)	(0, 0, 1, 0)	(0, 0, 0, 1)
1	(1, 0, 1, 0)	(0, 1, 0, 1)	(0, 0, 1, 0)
2	(0, 2, 0, 1)	(1, 0, 2, 0)	(0, 1, 0, 1)
3	(2, 0, 3, 0)	(0, 3, 0, 2)	(1, 0, 2, 0)
4	(0, 5, 0, 3)	(3, 0, 5, 0)	(0, 3, 0, 2)
5	(5, 0, 8, 0)	(0, 8, 0, 5)	(3, 0, 5, 0)
6	(0, 13, 0, 8)	(8, 0, 13, 0)	(0, 8, 0, 5)
7	(13, 0, 21, 0)	(0, 21, 0, 13)	(8, 0, 13, 0)
8	(0, 34, 0, 21)	(21, 0, 34, 0)	(0, 21, 0, 13)
9	(34, 0, 55, 0)	(0, 55, 0, 34)	(21, 0, 34, 0)
10	(0, 89, 0, 55)	(55, 0, 89, 0)	(0, 55, 0, 34)

© Springer Nature Switzerland AG 2019
S. Ault and C. Kicey, *Counting Lattice Paths Using Fourier
Methods*, Applied and Numerical Harmonic Analysis,
https://doi.org/10.1007/978-3-030-26696-7

1.3 *Proof by induction.*

Proof First, it will clarify notation somewhat to let $d = h + 1$. Moreover, we need not mention the height in our notation because the height is fixed. Then we need to show the following.

$$v_{n,a}(k) = v_{n,d-a}(d - k) \tag{B.12}$$

Base Case: $v_{0,a}(k) = 1$ if $k = a$ and 0 if $k \neq a$. On the other hand, $v_{0,d-a}(d - k) = 1$ if $d - k = d - a$ and 0 otherwise. But $d - k = d - a \iff k = a$, so $v_{0,a}(k) = v_{0,d-a}(d - k)$.

Inductive Step: Let $n \geq 1$ and suppose $v_{n,a}(k) = v_{n,d-a}(d - k)$ for all $1 \leq k \leq d - 1$. Consider $v_{n+1,d-a}(d - k)$.

$$
\begin{aligned}
v_{n+1,d-a}(d - k) &= v_{n,d-a}(d - k - 1) + v_{n,d-a}(d - k + 1) \\
&= v_{n,d-a}(d - [k + 1]) + v_{n,d-a}(d - [k - 1]) \\
&= v_{n,a}(k + 1) + v_{n,a}(k - 1), \quad \textit{by inductive hypothesis} \\
&= v_{n+1,a}(k)
\end{aligned}
$$

Equation (B.12) follows by induction.

1.5

$$
\begin{aligned}
(a + b + c)^n = (a + [b + c])^n &= \sum_{k=0}^{n} \binom{n}{k} a^{n-k} (b + c)^k \\
&= \sum_{k=0}^{n} \binom{n}{k} a^{n-k} \sum_{\ell=0}^{k} \binom{k}{\ell} b^{k-\ell} c^\ell \\
&= \sum_{k=0}^{n} \sum_{\ell=0}^{k} \binom{n}{k}\binom{k}{\ell} a^{n-k} b^{k-\ell} c^\ell
\end{aligned}
$$

Therefore, we can expand:

$$(1 + x + x^2)^{10} = \sum_{k=0}^{10} \sum_{\ell=0}^{k} \binom{10}{k}\binom{k}{\ell} 1^{10-k} x^{k-\ell} (x^2)^\ell = \sum_{k=0}^{10} \sum_{\ell=0}^{k} \binom{10}{k}\binom{k}{\ell} x^{k+\ell}$$

The terms having $k + \ell = 5$ such that $\ell \leq k$ are displayed in the table below. The coefficient of x^5 in $(1 + x + x^2)^{10}$ is 1452.

k	ℓ	Coefficient $= \binom{10}{k}\binom{k}{\ell}$
5	0	$\binom{10}{5}\binom{5}{0} = 252$
4	1	$\binom{10}{4}\binom{4}{1} = 840$
3	2	$\binom{10}{3}\binom{3}{2} = 360$
	Sum:	1452

1.6

$$g(x) = \sum_{k \geq 0}(1)x^k = 1 + x + x^2 + x^3 + \cdots$$

Right Shift: $$xg(x) = \sum_{k \geq 0}(1)x^{k+1} = x + x^2 + x^3 + x^4 + \cdots$$

As long as $|x| < 1$, both series are absolutely convergent (verify this using the Ratio or Root Test, for example). Therefore the difference can be taken term-by-term, with most terms cancelling.

$$g(x) - xg(x) = (1 + x + x^2 + x^3 + \cdots) - (x + x^2 + x^3 + \cdots)$$
$$(1 - x)g(x) = 1$$
$$g(x) = \frac{1}{1 - x}$$

1.7

$$C_n = v_{2n}^{(\infty)}(1) = \left(\begin{array}{c} 2n \\ \frac{1}{2}(2n + 1 - 1) \end{array}\right) - \left(\begin{array}{c} 2n \\ \frac{1}{2}(2n + 1 + 1) \end{array}\right)$$

$$= \binom{2n}{n} - \binom{2n}{n + 1}$$

$$= \frac{(2n)!}{n!n!} - \frac{(2n)!}{(n + 1)!(n - 1)!}$$

$$= \frac{(2n)!(n + 1) - (2n)!n}{(n + 1)!n!}$$

$$= \frac{(2n)!}{(n + 1)n!n!} = \frac{1}{n + 1}\binom{2n}{n}$$

1.8 Height ≤ 5 means that the associated lattice path is in the corridor $\mathbb{N}_0 \times \mathcal{C}^{(5)}$. Size 6 means that there are 6 edges in the tree, each of which must be traversed in both directions, so $n = 12$. Finally, $a = k = 1$ because the traversal must begin and end at the root note. Therefore, the number of such Catalan trees is equal to $v_{12}^{(5)}(1) = 122$.

1.12 Expand $p_n(x) = (x - x^{-1})(x + x^{-1})^n$ using the Binomial Theorem.

$$\sum_{j=0}^{n} \binom{n}{j} x^{n-2j+1} - \sum_{j=0}^{n} \binom{n}{j} x^{n-2j-1} = \sum_{j=1}^{n+1} \binom{n}{j-1} x^{n-2j-1} - \sum_{j=0}^{n} \binom{n}{j} x^{n-2j-1}$$

Thus, the $k = n - 2j - 1$ coefficient is $\binom{n}{j-1} - \binom{n}{j}$. Solving for j, we obtain $j = \frac{1}{2}(n - k - 1)$, and so $p_{n,k} = \binom{n}{\frac{1}{2}(n-k-1)} - \binom{n}{\frac{1}{2}(n-k+1)} = v_n^{(\infty)}(k)$.

2.1 Let $w(x) = T[v](x) = v(x - 1) + v(x + 1)$. We must show that $w\rho = -w$ and $w\sigma = -w$, given that v is admissible.

$$w\rho(x) = w(-x) = v(-x - 1) + v(-x + 1) = -v(x + 1) - v(x - 1) = -w(x)$$

$$\begin{aligned}
w\sigma(x) = w(2d - x) &= v(2d - x - 1) + v(2d - x + 1) \\
&= v(2d - [x + 1]) + v(2d - [x - 1]) \\
&= -v(x + 1) - v(x - 1) = -w(x)
\end{aligned}$$

2.3 $v(N - x) = -v(-N + x)$ by antisymmetry. Then by N-periodicity, $-v(-N + x) = -v(x)$.

2.4

(a) $e^{-\frac{2\pi i}{4} x\omega} = i^{x\omega}$. Thus,

$$U(\omega) = (0)i^{0\omega} + (1)i^{1\omega} + (0)i^{2\omega} + (-1)i^{3\omega} = i^{\omega} - (-i)^{\omega}$$

Therefore, $U = (\ldots, \mathbf{0}, 2i, 0, -2i, \ldots)$.

(b) $e^{-\frac{2\pi i}{6} x\omega} = e^{-\frac{\pi i}{3} x\omega}$. Using the period $x = -2, -1, 0, 1, 2, 3$ for the DFT,

$$U(\omega) = 1e^{\frac{2\pi i}{3}\omega} - 3e^{\frac{\pi i}{3}\omega} + 3e^{\frac{-\pi i}{3}\omega} - 1e^{-\frac{2\pi i}{3}\omega} = 2i\left[\sin\left(\frac{2\pi\omega}{3}\right) - 3\sin\left(\frac{\pi\omega}{3}\right)\right]$$

As a 6-periodic sequene, $U = (\ldots, \mathbf{0}, -2\sqrt{3}i, -4\sqrt{3}i, 0, 4\sqrt{3}i, 2\sqrt{3}i, \ldots)$.

(c) $e^{-\frac{2\pi i}{8}x\omega} = e^{-\frac{\pi i}{4}x\omega}$. Using the period $x = -3, -2, \ldots, 4$ for the DFT,

$$U(\omega) = e^{\frac{3\pi i}{4}\omega} + e^{\frac{\pi i}{4}\omega} + 3 + e^{\frac{-\pi i}{4}\omega} + e^{-\frac{3\pi i}{4}\omega} + 2e^{-\pi i\omega}$$

$$= 3 + 2(-1)^\omega + 2\left[\cos\left(\frac{3\pi\omega}{4}\right) + \cos\left(\frac{\pi\omega}{4}\right)\right]$$

Therefore, $U = (\ldots, 9, 1, 5, 1, 1, 1, 5, 1, \ldots)$.

(d) $U(\omega) = 5 + 2e^{\frac{-2\pi i}{3}\omega} + 3e^{-\pi i\omega} = 5 + 2\cos(\frac{2\pi\omega}{3}) + 3(-1)^\omega - \left(2\sin(\frac{2\pi\omega}{3})\right)i$.
$(U(\omega))_{\omega=0}^5 = \left(10, 1 - \sqrt{3}i, 7 + \sqrt{3}i, 4, 7 - \sqrt{3}i, 1 + \sqrt{3}i\right)$.

2.6

(a) Let $U(\omega) = \mathcal{F}[u](\omega) = \sum_{y=0}^{d-1} u(y)e^{-\frac{2\pi i}{N}y\omega}$. Then for $x \in \mathbb{Z}$ such that $0 \leq x \leq N - 1$, we have

$$\mathcal{F}^{-1}[U](x) = \frac{1}{N}\sum_{\omega=0}^{N-1} U(\omega)e^{\frac{2\pi i}{N}\omega x} = \frac{1}{N}\sum_{\omega=0}^{N-1}\left(\sum_{y=0}^{N-1} u(y)e^{-\frac{2\pi i}{N}y\omega}\right)e^{\frac{2\pi i}{N}\omega x}$$

$$= \frac{1}{N}\sum_{\omega=0}^{N-1}\sum_{y=0}^{N-1} u(y)e^{-\frac{2\pi i}{N}\omega(x-y)}$$

$$= \frac{1}{N}\sum_{y=0}^{N-1} u(y)\sum_{\omega=0}^{N-1} e^{-\frac{2\pi i}{N}\omega(x-y)}$$

The series $\sum_{\omega=0}^{N-1} e^{-\frac{2\pi i}{N}\omega(x-y)}$ sums to zero unless $x - y = 0$, i.e., $x = y$ when its sum is N (Eq. (A.9)), so the above simplifies to $\frac{1}{N}u(x) \cdot N = u(x)$.

(b) Similar to part (a).

2.7 With $x_0 = \omega_0 = 0$, $\sum |u(x)|^2 = 2^2 + 3^2 + 1^2 + (-4)^2 = 30$, and $\sum |U(\omega)|^2 = 2^2 + \left(\sqrt{1^2 + (-7)^2}\right)^2 + 4^2 + \left(\sqrt{1^2 + 7^2}\right)^2 = 120$, which when divided by $N = 4$ yields 30.

2.10

(a)

$$a(-x) = \frac{1}{2}[u(-x) + u(x)] = a(x)$$

$$b(-x) = \frac{1}{2}[u(-x) - u(x)] = -\frac{1}{2}[-u(-x) + u(x)] = -b(x)$$

Moreover, $a(x) + b(x) = \frac{1}{2}[2u(x)] = u(x)$.

(b) *Hints:* Because u is N-periodic, so are a and b; hence it suffices to take $x_0 = 0$. Prove the result first for real sequences (so that $\overline{b(x)} = b(x)$), then for the general case, break into real and imaginary parts.

(c) It may help to write $u = (\dots, 5, 0, 2, 3, 0, 0, \mathbf{5}, 0, 2, 3, 0, 0, 5, \dots)$. Then on the period $-2, -1, \dots, 3$,

$$a = \frac{1}{2}[(0, 0, \mathbf{5}, 0, 2, 3) + (2, 0, \mathbf{5}, 0, 0, 3)] = (1, 0, \mathbf{5}, 0, 1, 3)$$

$$b = \frac{1}{2}[(0, 0, \mathbf{5}, 0, 2, 3) - (2, 0, \mathbf{5}, 0, 0, 3)] = (-1, 0, \mathbf{0}, 0, 1, 0)$$

With $d = 3$ ($N = 2(3) = 6$), and by (2.41), $\mathcal{F}[a](\omega) = 5 + 3(-1)^\omega + 2\cos(\frac{2\pi\omega}{3})$. Then by (2.42), $\mathcal{F}[b](\omega) = -2i\sin(\frac{2\pi\omega}{3})$. By linearity, $\mathcal{F}[u] = \mathcal{F}[a] + \mathcal{F}[b]$, which matches the solution to Exercise 2.4(d).

2.11

(a) $\displaystyle v_{n,1}^{(5)}(x) = \frac{2^n}{6}\sum_{\omega=0}^{11}\sin\left(\frac{\pi\omega x}{6}\right)\cos^n\left(\frac{\pi\omega}{6}\right)\sin\left(\frac{\pi\omega}{6}\right)$

(b) $\displaystyle v_{n,2}^{(4)}(x) = \frac{2^n}{5}\sum_{\omega=0}^{9}\sin\left(\frac{\pi\omega x}{5}\right)\cos^n\left(\frac{\pi\omega}{5}\right)\sin\left(\frac{2\pi\omega}{5}\right)$

(c) $\displaystyle m_{n,1}^{(5)}(x) = \frac{1}{6}\sum_{\omega=0}^{11}\sin\left(\frac{\pi\omega x}{6}\right)\left[1 + 2\cos\left(\frac{\pi\omega}{6}\right)\right]^n\sin\left(\frac{\pi\omega}{6}\right)$

2.12

$$T_M^n = (I + [R + L])^n = \sum_{k=0}^{n}\binom{n}{k}I^{n-k}[R + L]^n = \sum_{k=0}^{n}\binom{n}{k}T^n$$

Equation (2.32) follows immediately.

2.15 First observe that $g_{n,a}(t)$ is even, so we may write:

$$c_{n,a}^{(d)} = \frac{2^n}{2\pi}\int_{-\pi}^{\pi}(1 + \cos(t))\cos^n(t)\frac{\sin(t[2j + 1])}{\sin(t)}\,dt$$

Then using Exercise A.10, the formula simplifies.

$$c^{(\infty)}_{2m,2j+1} = \frac{2^{2m}}{2\pi} \int_{-\pi}^{\pi} (1 + \cos(t)) \cos^{2m}(t) \left[\sum_{k=-j}^{j} e^{2kti} \right] dt$$

$$= \frac{2^{2m}}{2\pi} \sum_{k=-j}^{j} \int_{-\pi}^{\pi} \left(1 + \frac{1}{2}\left[e^{ti} + e^{-ti}\right]\right) \cdot \frac{1}{2^{2m}} \left(e^{ti} + e^{-ti}\right)^{2m} e^{2kti} \, dt$$

$$= \frac{1}{4\pi} \sum_{k=-j}^{j} \int_{-\pi}^{\pi} \left(2 + e^{ti} + e^{-ti}\right) \cdot \left(e^{ti} + e^{-ti}\right)^{2m} e^{2kti} \, dt$$

$$= \frac{1}{4\pi} \sum_{k=-j}^{j} \int_{-\pi}^{\pi} \left(2 + e^{ti} + e^{-ti}\right) \left[\sum_{\ell=0}^{2m} \binom{2m}{\ell} e^{\ell ti} e^{-(2m-\ell)ti} \right] e^{2kti} \, dt$$

$$= \frac{1}{4\pi} \sum_{k=-j}^{j} \sum_{\ell=0}^{2m} \left[\binom{2m}{\ell} \int_{-\pi}^{\pi} \left(2 + e^{ti} + e^{-ti}\right) e^{2\ell ti - 2mti + 2kti} \, dt \right]$$

Now we may use the integral formula (A.7) for complex exponentials. Note that the only nonzero integral occurs when $2\ell ti - 2mti + 2kti = 0$, or equivalently, $\ell = m - k$. Furthermore, when that situation occurs, the integral evaluates to: $\int_{-\pi}^{\pi} 2e^0 \, dt = 4\pi$. Therefore, we obtain:

$$c^{(\infty)}_{2m,2j+1} = \frac{1}{4\pi} \sum_{k=-j}^{j} \binom{2m}{m-k} 4\pi = \sum_{k=-j}^{j} \binom{2m}{m-k}$$

The three remaining cases for the general formula follow from the first with straightforward lattice path counting arguments and properties of Pascal's triangle. We will work out the case in which both parameters are odd. Note that when the initial point of paths is odd, then each even length path comes from two distinct odd length paths in the infinite corridor.

$$c^{(\infty)}_{2m+1,\,2j+1} = \frac{1}{2} c^{(\infty)}_{2m+2,\,2j+1}$$

$$= \frac{1}{2} \sum_{k=-j}^{j} \binom{2m+2}{m-k+1}$$

$$= \frac{1}{2} \sum_{k=-j}^{j} \left[\binom{2m+1}{m-k} + \binom{2m+1}{m-k+1} \right]$$

$$= \frac{1}{2} \left[\binom{2m+1}{m+j+1} + 2\left(\binom{2m+1}{m+j} + \cdots + \binom{2m+1}{m-j+1}\right) + \binom{2m+1}{m-j} \right]$$

$$= \binom{2m+1}{m+j} + \binom{2m+1}{m+j-1} + \cdots + \binom{2m+1}{m-j+1} + \binom{2m+1}{m-j}$$

$$= \sum_{k=-j}^{j} \binom{2m+1}{m-k}$$

The next to last line follows because $(m + j - 1) + (m - j) = 2m - 1$ and so the first and last binomials have the same value.

In both cases above, we have $c_{n,2j+1}^{(\infty)} = \sum_{k=-j}^{j} \binom{n}{\lfloor n/2 \rfloor - k}$. The cases $c_{n,2j}$ are handled analogously.

3.1 Nonidentity moves: $3^r - 1$. By Lemma 3.3, each balanced move set corresponds to a transition operator of the form $\sum_{\mathbf{k} \in \{0,1\}^r} b_{\mathbf{k}} T_1^{k_1} \cdots T_r^{k_r}$, where each $b_{\mathbf{k}}$ can be either 1 or 0. There are 2^r distinct monomials $T_1^{k_1} \cdots T_r^{k_r}$, and the choices of $b_{\mathbf{k}}$ are equivalent to choosing a subset of the set of monomials. However, the *empty* subset (in which every $b_{\mathbf{k}}$ is zero) does not correspond to a move set. Therefore, the number of balanced move sets is $2^{2^r} - 1$.

3.3 Let $v(\mathbf{x})$ be a function, and let $w(\mathbf{x}) = v(\gamma^{-1}\mathbf{x})$. Noting that $w = H^\gamma[v]$, we find:

$$H^{\eta\gamma}[v](\mathbf{x}) = v((\eta\gamma)^{-1}\mathbf{x}) = v(\gamma^{-1}\eta^{-1}\mathbf{x}) = w(\eta^{-1}\mathbf{x}) = H^\eta[w](\mathbf{x}) = H^\eta\left[H^\gamma[v]\right](\mathbf{x})$$

3.4

$$H^{\rho_k}L_j = \begin{cases} L_j H^{\rho_k} & k \neq j, \\ R_j H^{\rho_k} & k = j \end{cases} \quad \text{and} \quad H^{\sigma_k}L_j = \begin{cases} L_j H^{\sigma_k} & k \neq j, \\ R_j H^{\sigma_k} & k = j \end{cases}$$

Now let $\gamma = \rho_k$ or σ_k. If $k \neq j$, then H^γ commutes with both R_j and L_j, and so $H^\gamma T_j = T_j H^\gamma$ follows. Moreover if $k = j$, then we have:

$$H^\gamma T_j = H^\gamma R_j + H^\gamma L_j = L_j H^\gamma + R_j H^\gamma = T_j H^\gamma$$

3.6 Fix $\mathbf{a} \in \mathbb{Z}^r$ and suppose $\mathbf{x} \in \mathbb{Z}^r$ is arbitrary. Let $1 \leq j \leq r$.

$$\begin{aligned}
\delta_{\mathbf{a}}\rho_j(\mathbf{x}) &= \delta_{\mathbf{a}}(x_1, \ldots, -x_j, \ldots, x_r) \\
&= (x_1 - a_1, \ldots, -x_j - a_j, \ldots, x_r - a_r) \\
&= \begin{cases} 1 & \text{if } x_1 = a_1, \ldots, x_j = -a_j, \ldots, \text{ and } x_r = a_r, \\ 0 & \text{otherwise} \end{cases} \\
&= \delta_{(a_1, \ldots, -a_j, \ldots, a_r)}(\mathbf{x})
\end{aligned}$$

3.7

$$\Delta_{(1,2,3)} = \sum_{\mathbf{k} \in \{0,1\}^3} (-1)^{\sum k_j} \delta_{((-1)^{k_1}(1),\, (-1)^{k_2}(2),\, (-1)^{k_r}(3))}$$

$$\begin{aligned}
&= \delta_{(1,2,3)} - \delta_{(1,2,-3)} - \delta_{(1,-2,3)} + \delta_{(1,-2,-3)} \\
&\quad - \delta_{(-1,2,3)} + \delta_{(-1,2,-3)} + \delta_{(-1,-2,3)} - \delta_{(-1,-2,-3)}
\end{aligned}$$

3.10 *Hint:* Proof by induction.

3.11 Use (3.21) to find $\widehat{T}(\omega)$ in each case.

(a) $\widehat{T}(\omega) = 2\cos\left(\frac{\pi\omega_2}{3}\right) + 4\cos\left(\frac{\pi\omega_1}{7}\right)\cos\left(\frac{\pi\omega_2}{3}\right)$

(b) $\widehat{T}(\omega) = 1 + 2\cos\left(\frac{\pi\omega_3}{8}\right) + 8\cos\left(\frac{\pi\omega_1}{4}\right)\cos\left(\frac{\pi\omega_2}{6}\right)\cos\left(\frac{\pi\omega_3}{8}\right)$

(c) $\widehat{T}(\omega) = 2\left[\cos\left(\frac{\pi\omega_1}{d_1}\right) + \cos\left(\frac{\pi\omega_2}{d_2}\right) + \cos\left(\frac{\pi\omega_3}{d_3}\right) + \cos\left(\frac{\pi\omega_4}{d_4}\right)\right]$

(d) $\widehat{T}(\omega) = 2\sum_{k=1}^{r}\cos\left(\frac{\pi\omega_k}{d_k}\right)$

3.12 If r is odd, then $(-2i)^r = -2^r(-1)^{\frac{r-1}{2}}i$. Apply the inverse DFT to (3.23), using Euler's Formula and the fact that V_n is purely imaginary.

$$v_n(\mathbf{x}) = \frac{-2^r(-1)^{\frac{r-1}{2}}i}{2^r\prod_{j=1}^r d_j}\sum_{\omega=-d+1}^{d} i\sin(\pi i\mathbf{x}\cdot(\omega/d))\left[\widehat{T}(\omega)\right]^n\prod_{j=1}^r\sin\left(\frac{\pi\omega_j a_j}{d_j}\right)$$

$$= \frac{(-1)^{\frac{r-1}{2}}}{\prod_{j=1}^r d_j}\sum_{\omega=-d+1}^{d}\sin(\pi i\mathbf{x}\cdot(\omega/d))\left[\widehat{T}(\omega)\right]^n\prod_{j=1}^r\sin\left(\frac{\pi\omega_j a_j}{d_j}\right)$$

4.4

v_0 :

1	

v_1 :

1	1
1	1

v_2 :

2	1
4	2
4	2

v_3 :

6	9
10	15
8	12

v_4 :

40	24
60	36
45	27

v_5 :

100	160
145	232
105	168

v_6 :

637	392
910	560
650	400

$\langle v_4, v_2\rangle = (2)(40) + (4)(60) + (4)(45) + (1)(24) + (2)(36) + (2)(27) = 650.$
$v_6(2,1) = 650.$

4.6

$$\mathcal{F}^{-1}\left[V_0^{(j)}(\omega)\right](x) = \mathcal{F}^{-1}\left[(-1)^{\frac{j+1}{2}}2i\,\Delta_j\right](x)$$

$$= (-1)^{\frac{j+1}{2}}2i\mathcal{F}^{-1}\left[\Delta_j\right](x)$$

$$= (-1)^{\frac{j+1}{2}}2i\cdot\frac{1}{4m}\left(e^{\frac{2\pi i}{4m}jx} - e^{\frac{2\pi i}{4m}(-j)x}\right)$$

$$= \frac{(-1)^{\frac{j+1}{2}}i}{2m}\left(e^{\frac{\pi i}{2m}jx} - e^{-\frac{\pi i}{4m}jx}\right)$$

$$= \frac{(-1)^{\frac{j+1}{2}}i}{2m}\cdot 2i\sin\left(\frac{\pi jx}{2m}\right)$$

$$= \frac{(-1)^{\frac{j-1}{2}}}{m}\sin\left(\frac{\pi jx}{2m}\right)$$

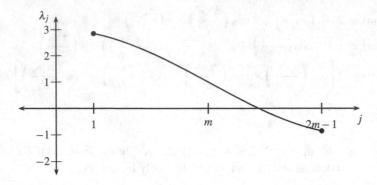

Fig. 1 Graph of $\lambda_j = 1 + 2\cos(\frac{j\pi}{2m})$ for $1 \le j \le 2m - 1$

4.11

(a) $\widehat{T}(\omega)^2 = 4\cos^2(\frac{\omega\pi}{8})$, so we have the following values at integers ω.

ω	-7	-6	-5	-4	-3	-2	-1	0	1	2	3	4	5	6	7	8
$\widehat{T}(\omega)^2$	$2+\sqrt{2}$	2	$2-\sqrt{2}$	0	$2-\sqrt{2}$	2	$2+\sqrt{2}$	4	$2+\sqrt{2}$	2	$2-\sqrt{2}$	0	$2-\sqrt{2}$	2	$2+\sqrt{2}$	4

Noting that $W^{(1)} = V^{(1)} + V^{(7)}$ and $W^{(1)} = V^{(3)} + V^{(5)}$, the appropriate eigenvalues would be $\mu_1 = 2 + \sqrt{2}$ and $\mu_2 = 2 - \sqrt{2}$.

(b) $c_{2n} = \frac{1}{2}(1 + \sqrt{2})(2 + \sqrt{2})^n + \frac{1}{2}(1 - \sqrt{2})(2 - \sqrt{2})^n$.

4.12 *Hint:* See Fig. 1.

4.16 (a) $c_{2n}/4^n$ where $c_{2n} = 4 \cdot 3^{n-1}$. (b) $d_{2n}/4^n$ where $d_{2n} = \binom{2n}{n-1} + \binom{2n}{n} + \binom{2n}{n+1}$.

A.1

(a)

$$\frac{2}{1-e^{i\theta}} = \frac{2}{1-e^{i\theta}} \frac{\overline{1-e^{i\theta}}}{1-e^{i\theta}} = \frac{2}{1-e^{i\theta}} \frac{1-e^{-i\theta}}{1-e^{-i\theta}}$$

$$= 2\frac{1-e^{-i\theta}}{(1-e^{i\theta})(1-e^{-i\theta})} = 2\frac{1-[\cos(-\theta)+i\sin(-\theta)]}{2-2\cdot\frac{1}{2}(e^{i\theta}+e^{-i\theta})}$$

$$= \frac{1-\cos\theta+i\sin\theta}{1-\cos\theta} = 1 + \left(\frac{\sin\theta}{1-\cos\theta}\right)i$$

(b)

$$\frac{1}{1-e^{i\theta}} - \frac{1}{1-e^{-i\theta}} = \frac{e^{i\theta}-e^{-i\theta}}{(1-e^{i\theta})(1-e^{-i\theta})} = \frac{2i}{2}\cdot\frac{\frac{1}{2i}(e^{i\theta}-e^{-i\theta})}{1-\frac{1}{2}(e^{i\theta}+e^{-i\theta})} = \frac{i\sin\theta}{1-\cos\theta}$$

A.4 We have $r = |z| = \sqrt{1^2 + 1^2} = \sqrt{2}$, and so $1 = r\cos\theta = \sqrt{2}\cos\theta \implies$
$\cos\theta = \frac{1}{\sqrt{2}}$. Similarly, $\sin\theta = \frac{1}{\sqrt{2}}$. Together these imply that $\theta = \frac{\pi}{4}$. Therefore,
$z = \sqrt{2}e^{\frac{i\pi}{4}}$ in polar form. So $z^2 = 2e^{\frac{i\pi}{2}}$, $z^3 = 2\sqrt{2}e^{\frac{i(3\pi)}{4}}$, and $z^4 = 4e^{i\pi}$.

The sequence of powers z^0, z^1, z^2, z^3, z^4 would start with the vector $1 = 1 + 0i$ and proceed counterclockwise, to move half way around the unit circle in angular integer multiples of $\pi/4$, scaling length by a factor of $\sqrt{2}$ each time, as illustrated below.

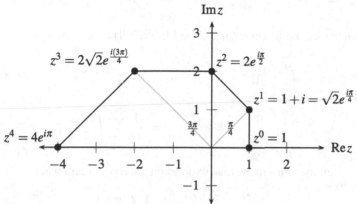

A.5 Using the sum of the geometric series and Exercise A.1(b), we have

$$\sum_{k=1}^{7} \sin\left(\frac{k\pi}{8}\right) = \sum_{k=0}^{7} \sin\left(\frac{k\pi}{8}\right) = \frac{1}{2i}\sum_{k=0}^{7}\left[\left(e^{i\frac{\pi}{8}}\right)^k - \left(e^{-i\frac{\pi}{8}}\right)^k\right]$$

$$= \frac{1}{2i}\left[\frac{1 - e^{i\pi}}{1 - e^{i\frac{\pi}{8}}} - \frac{1 - e^{-i\pi}}{1 - e^{-i\frac{\pi}{8}}}\right]$$

$$= \frac{1}{2i}\left[\frac{2}{1 - e^{i\frac{\pi}{8}}} - \frac{2}{1 - e^{-i\frac{\pi}{8}}}\right]$$

$$= \frac{\sin\left(\frac{\pi}{8}\right)}{1 - \cos\left(\frac{\pi}{8}\right)}$$

A.6 For $k = 0$, $\int_{-\pi}^{\pi} e^0\, d\theta = \pi - (-\pi) = 2\pi$. Now suppose $k \neq 0$.

$$\int_{-\pi}^{\pi} e^{ik\theta}\, d\theta = \int_{-\pi}^{\pi} (\cos(k\theta) + i\sin(k\theta))\, d\theta = \int_{-\pi}^{\pi} \cos(k\theta)\, d\theta + i\int_{-\pi}^{\pi} \sin(k\theta)\, d\theta = 0$$

For (A.8), use the substitution $u = \frac{2\pi}{d}\theta$. Then with $u_0 = \frac{2\pi\theta_0}{d}$, the integral becomes:

$$\int_{\theta_0}^{\theta_0 + d} e^{i\frac{2\pi}{d}k\theta}\, d\theta = \frac{d}{2\pi}\int_{u_0}^{u_0 + 2\pi} e^{iku}\, du$$

The result now follows from (A.7).

A.7 First, if $k = 0$, then $\sum_{\omega=0}^{d-1} e^{i\frac{2k\pi}{d}\omega} = \sum_{\omega=0}^{d-1} 1 = d$. Next, suppose $k \neq 0$ is an integer.

$$\sum_{\omega=0}^{d-1} \left(e^{i\frac{2k\pi}{d}}\right)^{\omega} = \frac{1 - \left(e^{i\frac{2k\pi}{d}}\right)^d}{1 - e^{i\frac{2k\pi}{d}}} = \frac{1 - e^{i(2k\pi)}}{1 - e^{i\frac{2k\pi}{d}}} = \frac{1 - 1}{1 - e^{i\frac{2k\pi}{d}}} = 0$$

A.8 From the Binomial Theorem and De Moivre's Theorem, we have

$$\int_{-\pi}^{\pi} \cos^{2m}(t)\, dt = \int_{-\pi}^{\pi} \left[\frac{1}{2}(e^{it} + e^{-it})\right]^{2m} dt = \frac{1}{4^m} \int_{-\pi}^{\pi} \sum_{k=0}^{2m} \binom{2m}{k} e^{it(2m-k)} e^{-itk}\, dt$$

$$= \frac{1}{4^m} \sum_{k=0}^{2m} \binom{2m}{k} \int_{-\pi}^{\pi} e^{it(2m-2k)}\, dt$$

By (A.7), all the terms in the final expression are zero, except when $2m - 2k = 0$ or $k = m$. Thus:

$$\int_{-\pi}^{\pi} \cos^{2m}(t)\, dt = \frac{1}{4^m} \binom{2m}{m}$$

This quantity is the ratio of the weight of central entries on Pascal's triangle to the total weight on the entire row.

A.9 Rewrite the integrand using Euler's Formula.

$$\prod_{k=1}^{m} \cos(kx) = \prod_{k=1}^{m} \frac{e^{ikx} + e^{-ikx}}{2} = \frac{1}{2^m} \sum_{\varepsilon_k = \pm 1} e^{i(\varepsilon_1 + 2\varepsilon_2 + \cdots + m\varepsilon_m)x}$$

Now by (A.7), $\int_0^{2\pi} e^{i(\varepsilon_1 + 2\varepsilon_2 + \cdots + m\varepsilon_m)x}\, dx \neq 0$ if and only if $\varepsilon_1 + 2\varepsilon_2 + \cdots + m\varepsilon_m = 0$, in which case the value is 2π. Therefore, $I_m \neq 0$ as long as there is at least one way to express $\varepsilon_1 + 2\varepsilon_2 + \cdots + m\varepsilon_m = 0$ by choices of $\varepsilon_k \in \{-1, 1\}$.

Suppose $\varepsilon_1 + 2\varepsilon_2 + \cdots + m\varepsilon_m = 0$. Then the sum $1 + 2 + \cdots + m$ must be even (reduce modulo 2, and observe that $-1 \equiv 1 \bmod 2$). But since $1 + 2 + \cdots + m = \frac{1}{2}m(m + 1)$, this implies that $m(m + 1)$ is a multiple of 4. This in turn forces either m or $m + 1$ to be a multiple of 4. Thus, $I_m \neq 0$ if and only if $m = 4\ell$ or $m = 4\ell - 1$ for some $\ell \in \mathbb{N}$.

B.2

(a) $\left(\frac{1}{2}, \frac{\sqrt{3}}{2}\right)$ (b) $\left(-\frac{9}{2}, \frac{5\sqrt{3}}{2}\right)$ (c) $\left(12, -12\sqrt{3}\right)$ (d) $\left(\frac{d}{2}, \frac{d\sqrt{3}}{2}\right)$

B.3 For $\Delta(4, 4, 2)$:

$$M_1 = \begin{pmatrix} 1 & 0 \\ 0 & -1 \end{pmatrix}, \quad M_2 = \begin{pmatrix} 0 & 1 \\ 1 & 0 \end{pmatrix}, \quad \rho_3 \begin{pmatrix} a \\ b \end{pmatrix} = \begin{pmatrix} 0 & -1 \\ -1 & 0 \end{pmatrix} \begin{pmatrix} a \\ b \end{pmatrix} + \begin{pmatrix} d \\ d \end{pmatrix}$$

For $\Delta(4, 2, 4)$:

$$M_1 = \begin{pmatrix} 1 & 0 \\ 0 & -1 \end{pmatrix}, \quad M_2 = \begin{pmatrix} 0 & 1 \\ 1 & 0 \end{pmatrix}, \quad \rho_3 \begin{pmatrix} a \\ b \end{pmatrix} = \begin{pmatrix} -1 & 0 \\ 0 & 1 \end{pmatrix} \begin{pmatrix} a \\ b \end{pmatrix} + \begin{pmatrix} 2d \\ 0 \end{pmatrix}$$

B.7 Because $\mathbf{x} \in \ell$, we have $\gamma(\mathbf{x}) = \mathbf{x}$. Then $v(\mathbf{x}) = v(\gamma^{-1}\mathbf{x}) = H^\gamma[v](\mathbf{x}) = -v(\mathbf{x})$, implying that $v(\mathbf{x}) = 0$.

References

1. L. Addario-berry, B.A. Reed, *Ballot Theorems, Old and New*. Bolyai Society Mathematical Studies (Series) (2008), pp. 9–35
2. A.V. Aho, J.E. Hopcroft, J.D. Ullman, *The Design and Analysis of Computer Algorithms* (Addison-Wesley, Boston, 1974)
3. D. André, Solution directe du problème résolu par M Bertrand. Comptes Rendus Acad. Sci. Paris **105**, 436–437 (1887)
4. George E. Andrews, Some formulae for the Fibonacci sequence with generalizations. Fibonacci Quart. **7**(2), 113–130 (1969)
5. M.A. Amstrong, *Groups and Symmetry Undergraduate Texts in Mathematics* (Springer-Verlag, New York, Berlin, Heidelberg, 1988)
6. S.V. Ault, C. Kicey, Counting paths in corridors using circular Pascal arrays. Discret. Math. **332**(6), 45–54 (2014)
7. C. Banderier, P. Flajolet, Basic analytic combinatorics of directed lattice paths. Theor. Comput. Sci. **281**(1–2), 37–80 (2002)
8. C. Banderier, P. Nicodème, Bounded discrete walks, in *DMTCS Proceedings, 21th International Meeting on Probabilistic, Combinatorial, and Asymptotic Methods in the Analysis of Algorithms (AofA'10)*, pp. 35–48 (2010)
9. C. Banderier, M. Wallner, Lattice paths of slope 2/5, in *Proceedings of the Twelfth Workshop on Analytic Algorithmics and Combinatorics (ANALCO)* (2015), pp. 105–113
10. J. Bertrand, Solution d'un problème. Comptes Rendus l'Académie Sci. **105**, 369 (1887)
11. R.S. Borden, *A Course in Advanced Calculus (Dover Books on Mathematics)* (Dover Publications, 2012)
12. R. Brak, J. Essam, J. Osborn, A.L. Owczarek, A. Rechnitzer, Lattice paths and the constant term, in *Journal of Physics: Conference Series*, vol. 42, no. (1), p. 47 (2006)
13. J.W. Brown, R.V. Churchill, *Complex Variables and Applications*, Brown and Churchill Series (McGraw-Hill Higher Education, Boston, 2009)
14. R.A. Brualdi, *Introductory Combinatorics*, Fifth edn. (Pearson, 2008)
15. C. Coker, Enumerating a class of lattice paths. Discr. Math. **271**, 13–28 (2003)
16. H.S.M. Coxeter, Discrete groups generated by reflections. Ann. Math. **35**, 588–621 (1934)
17. N. Dershowitz, C. Rinderknecht, The average height of Catalan trees by counting lattice paths. Math. Mag. **88**(3), 187–195 (2015)
18. E. Deutsch, Dyck path enumeration. Discr. Math. **204**, 167–202 (1999)

© Springer Nature Switzerland AG 2019
S. Ault and C. Kicey, *Counting Lattice Paths Using Fourier Methods*, Applied and Numerical Harmonic Analysis,
https://doi.org/10.1007/978-3-030-26696-7

19. P.G. Doyle, J.L. Snell, *Random Walks and Electric Networks* (The Mathematical Association of America, Washington, DC, 1984)
20. D.S. Dummit, R.M. Foote, *Abstract Algebra*, 3rd edn. (Wiley, Hoboken, 2004)
21. H. Dym, H.P. McKean, *Fourier Series and Integrals* (Academic Press, San Diego, 1972)
22. S. Elizalde, M. Rubey, Bijections for lattice paths between two boundaries, in *FPSAC 2012, Nagoya, Japan, DMTCS Proceedings AR*, pp. 827–838 (2012)
23. S.S. Epp, *Discrete Mathematics with Applications* (Cengage Learning, 2010)
24. W. Feller, *An Introduction to Probability Theory and its Applications*, vol. I, 3rd edn. (Wiley, New York, 1968)
25. S. Felsner, D. Heldt. Lattice path enumeration and Toeplitz matrices. J. Integer Seq. **18**, 15 (2015). Article 15.1.3
26. P. Flajolet, R. Sedgewick, *Analytic Combinatorics*, 1st edn. (Cambridge University Press, New York, 2009)
27. I.M. Gessel, C. Krattenthaler, Cyclic partitions. Trans. Amer. Math. Soc. **349**, 429–479 (1997)
28. I.M. Gessel, D. Zeilberger, Random walk in a Weyl chamber. Proc. Am. Math. Soc. **115**, 27–31 (1992)
29. D. Grabiner, Random walk in an alcove of an affine Weyl group, and non-colliding random walks on an interval. J. Comb. Theor. Ser. A **97**(2), 285–306 (2002)
30. R.K. Guy, C. Krattenthaler, B.E. Sagan, Lattice paths, reflections, and dimension-changing bijections. Ars Combin. **34**, 3–15 (1992)
31. J. Harris, J.L. Hirst, M. Mossinghoff, *Combinatorics and Graph Theory: Undergraduate Texts in Mathematics* (Springer, New York, 2008)
32. J.E. Humphreys, *Reflection Groups and Coxeter Groups* (Cambridge University Press, 1992)
33. K. Humphreys, A history and a survey of lattice path enumeration. J. Stat. Plan. Inference **140**(8), 2237–2254 (2010)
34. K.S. Kedlaya, B. Poonen, R. Vakil, *The William Lowell Putnam Mathematical Competition 1985–2000: Problems, Solutions, and Commentary* (Mathematical Association of America, Washington, DC, 2008)
35. C. Kicey, K. Klimko, Some geometry of Pascal's triangle. Pi Mu Epsil. J. **13**(4), 229–245 (2011)
36. C. Krattenthaler, Lattice path enumeration, in *Handbook of Enumerative Combinatorics*, ed. by M. Bóna (Discrete Mathematics and its Application) (CRC Press, New York, 2015)
37. C. Krattenthaler, S.G. Mohanty, Lattice path combinatorics—applications to probability and statistics, in *Encyclopedia of Statistical Sciences*, Second edn., ed. by N.L. Johnson, C.B. Read, N. Balakrishnan, B. Vidakovic (Wiley, New York, 2003)
38. C. Krattenthaler, S.G. Mohanty, On lattice path counting by major index and descents. Eur. J. Comb. **14**(1), 43–51 (1993)
39. D.C. Lay, S.R. Lay, J.J. McDonald, *Linear Algebra and its Applications*, Fifth edn (Pearson, 2015)
40. S.J. Leon, *Linear Algebra with Applications: Featured Titles for Linear Algebra* (Pearson, 2014)
41. W. Magnus, *Noneuclidean Tesselations and Their Groups (Pure and Applied Mathematics)* (Elsevier Science, 1974)
42. S.G. Mohanty, A short proof of Steck's result on two-sample Smirnov statistics. Ann. Math. Stat. **42**, 413–414 (1971)
43. S.G. Mohanty, *Lattice Path Counting and Applications* (Academic Press [Harcourt Brace Jovanovich], New York, 1979)
44. P.R.G. Mortimer, T. Prellberg, On the number of walks in a triangular domain. Preprint
45. T.V. Narayana, *Lattice Path Combinatorics with Statistical Applications* (Toronto University Press, Toronto, 1979)
46. D. Percival, A. Walden, *Wavelet Methods for Time Series Analysis* (Cambridge University Press, Cambridge, UK, 2000)
47. G. Pólya, Über eine aufgabe betreffend die irrfahrt im strassennetz. Math. Ann. **84**, 149–160 (1921)

48. D. Poole, R. Lipsett, *Linear Algebra: A Modern Introduction* (Cengage Learning, 2014)
49. V. Reiner, D. Stanton, D. White, Cyclic seiving. Not. AMS **61**(2), 169–171 (2014)
50. J. Rotman, *An Introduction to the Theory of Groups*, Graduate Texts in Mathematics (Springer, New York, 1999)
51. W. Rudin, *Principles of Mathematical Analysis*. International Series in Pure and Applied Mathematics. (Mcgraw-Hill, 1976)
52. E.B. Saff, A.D. Snider, *Fundamentals of Complex Analysis for Mathematics, Science and Engineering* (Prentice-Hall, Incorporated, 1976)
53. R. Sedgewick, P. Flajolet, *An Introduction to the Analysis of Algorithms* (Addison-Wesley Longman Publishing Co. Inc, Boston, 1996)
54. N.J.A. Sloane, The on-line encyclopedia of integer sequences. Published electronically at http://oeis.org/
55. J.O. Smith, *Mathematics of the Discrete Fourier Transform (DFT): with Audio Applications* (W3K Publishing. BookSurge Publishing, 2007)
56. M.Z. Spivey, Enumerating lattice paths touching or crossing the diagonal at a given number of lattice points. Electron. J. Comb. **19**(3), 1–6 (2012)
57. R.P. Stanley, *Enumerative Combinatorics*, vol. 1, 2nd edn. (Cambridge University Press, New York, 2011)
58. G.P. Steck, The Smirnov tests as rank tests. Ann. Math. Stat. **40**, 1449–1466 (1969)
59. G.P. Steck, Evaluation of some Steck determinants with applications. Comm. Stat. **3**, 121–138 (1974)
60. J. Stewart, *Calculus—Early Transcendentals*, 6th edn. (Thomson Brooks/Cole, Belmont, 2008)
61. Lajos Takács, Ballot problems. Z. Wahrscheinlichkeitstheorie Verwandte Geb. **1**(2), 154–158 (1962)
62. P. Tetali, Random walks and the effective resistance of networks. J. Theor. Probab. **4**, 101–109 (1991)
63. G.B. Thomas, M.D. Weir, J. Hass, F.R. Giordano, *Thomas' Calculus, Early Transcendentals*, Twelfth edn. (Addison-Wesley, Boston, 2010)
64. H.S. Wilf, *Generatingfunctionology* (Academic Press, 1994)

Index

A
Admissible, 24, 48, 105
Antisymmetry, 12, 24

B
Balanced, 51, 106
Ballot Problem, 10, 38
Binomial coefficient, 8
   central, 10, 38
Binomial Theorem, 9

C
Catalan number, 6, 19, 38
Catalan tree, 19
Complex
   conjugate, 90
   exponential, 91
   number, 89
   plane, 89
Corridor, 2, 46, 104
   centered, 74, 81
   cyclic, 38
   cylindrical, 38
   dual structure, 12
   infinite, 10
   number, 3
   path, 2
   skewed-bottom, 16
   skewed-top, 21
   toroidal, 63
   unrestricted, 10
Cycle graph, 38

D
Delta function, 30, 55, 74
Dirichlet kernel, 43, 96
Discrete Fourier Transform (DFT), *see*
   Fourier transform, discrete
Dot product, 94
Drunkard's walk, 3, 42
Dual corridor, 25
Dyck path, 5

E
Eigenspace, 74, 95
Eigenvalue, 73, 95
Eigenvector, 73, 95
Euler's Formula, 91
Euler's Identities, 91
Even, 41

F
Fibonacci numbers, 5, 21, 34, 42, 113
Floor, 10
Fourier transform, 28
   discrete, 28, 55
   inverse, 28, 55
   multidimensional, 54
Function space, 25
Fundamental
   corridor, 25
   region, 2, 25, 46, 104
   tile, 104

G
Generating function, 10

© Springer Nature Switzerland AG 2019
S. Ault and C. Kicey, *Counting Lattice Paths Using Fourier
Methods*, Applied and Numerical Harmonic Analysis,
https://doi.org/10.1007/978-3-030-26696-7